Physics of Self-Organization Systems

Physics of Self-Organization Systems

Proceedings of the 5th 21st Century COE Symposium

Tokyo, Japan 13 – 14 September 2007

editors

Shin'ichi Ishiwata

Yasushi Matsunaga

Waseda University, Japan

World Scientific

NEW JERSEY • LONDON • SINGAPORE • BEIJING • SHANGHAI • HONG KONG • TAIPEI • CHENNAI

Published by

World Scientific Publishing Co. Pte. Ltd.
5 Toh Tuck Link, Singapore 596224
USA office: 27 Warren Street, Suite 401-402, Hackensack, NJ 07601
UK office: 57 Shelton Street, Covent Garden, London WC2H 9HE

British Library Cataloguing-in-Publication Data
A catalogue record for this book is available from the British Library.

PHYSICS OF SELF-ORGANIZATION SYSTEMS
Proceedings of the 5th 21st Century COE Symposium

ISBN-13 978-981-279-336-2
ISBN-10 981-279-336-4

Printed in Singapore by World Scientific Printers

PREFACE

Various phenomena and structures observed in the biological world, materials, and space, can be considered as one of the aspects that self-organization systems exhibit as a result of the interplay between various elements (multi-components). Physics of the 20th century focused on simple systems composed of a single or a few types of components, but in this 21st Century COE (Center of Excellence) program conducted since 2003, we challenged the new objectives of physics, i.e., "Self-organizing systems composed of multi-components", in order to create a new field and establish universal comprehension in physics. The typical example of such a system is a biological system. Even though it is a living organism, it is also a "product" created by nature. It may be said therefore that a living being is a enthralling research object for the physics of the 21st century. It is a fascinating goal for physicists to reveal the mechanism, structure, and organization of biological systems from the viewpoint of physics, and to solve the puzzle of a biological system as being a physical one, just the same as the physics of space and materials. Even more so, as their components are created in the same environment, there must exist a common law that governs living organisms and materials, and materials and space.

Therefore, not only did we study these three apparently separate fields, but also postulated that living organisms, materials, and space have strong mutual links and it is worth studying them unitedly, as a single whole, or "matter". It is the long-standing tradition of Waseda University to emphasize the practical approach to science. However, in this COE program, we seek to come up with a new dimension of physics as a basic science. While taking advantage of the Waseda tradition to prioritize the applied science, we aimed to establish a research group that will be able to contribute to the development of basic science.

In particular, we committed ourselves to the education of young talented researchers. In order to venture into unknown realms of physics and explore frontier fields, the enthusiasm of young researchers is indispensable. The Major in Pure and Applied Physics at Waseda University fully has the potential to produce the next-generation scientists.

As the valuable activity of our 21st Century COE program, every year we organized an International Symposium. This book is the Proceedings of the last Symposium (the *Fifth International Symposium of our 21st Century COE program entitled "Physics toward the next generations"*) held at the Waseda International Conference Center on September 13-14, 2007. The Proceedings consist of three parts entitled Biophysics, Nonequilibrium Statistical Physics and Related Topics, and Astrophysics as Interdisciplinary Science. In each part, there are several articles written by young researchers, including not only post-doctoral fellows, but doctoral course students as well, from which, we hope, readers can recognize the high level of research activities done by members and students of our 21st Century COE Program, and by researchers collaborating with us.

How our program was successful should be evaluated not only from the research results presented in these Proceedings, but also from the education and research activities performed through this 21st Century COE Program. We sincerely hope that our endeavor, spanning the last several years to create the foundation for the high-level activities of the Major in Pure and Applied Physics at Waseda University, will be rewarded in future.

And finally but not the least, on behalf of the Local Organizing and Program Committees, we would like to express our sincere thanks to all the contributors and the participants of this Symposium.

Shin'ichi Ishiwata
(21COE Leader,
Professor of Major in Pure and Applied Physics)

Waseda University, Japan
31 January 2008

Yasushi Matsunaga
(21COE Manager, Visiting Associate Professor
in Faculty of Science and Engineering)

ORGANIZING COMMITTEES

S. Ishiwata (Chairman)	– Waseda University, Tokyo, Japan
K. Maeda	– Waseda University, Tokyo, Japan
C. Oshima	– Waseda University, Tokyo, Japan
Y. Uesu	– Waseda University, Tokyo, Japan
A. Tackeuchi	– Waseda University, Tokyo, Japan
S. Tasaki	– Waseda University, Tokyo, Japan
I. Terasaki	– Waseda University, Tokyo, Japan
K. Kinosita Jr.	– Waseda University, Tokyo, Japan
Y. Matsumaga	– Waseda University, Tokyo, Japan

1. K. Maeda
2. S. Uchida
3. S. Ishiwata
4. Y. Uesu
5. Y. Oono
6. P. Gaspard
7. Y. Matsunaga
8. I. Terasaki
9. B.-L. Hu
10. S. Kurihara
11. H. Nakazato
12. S. Tasaki
13. T. Nakano
14. T. Sunaga
15. T. Katsufuji
16. Y. Yamasaki
17. T. Monnai
18. Y. Klein
19. A. Matsuda
20. Y. Takamizu
21. T. Suzuki
22. T. Daishido
23. S. Shibasaki
24. H. Nakamura

CONTENTS

PART A

Biophysics

BIO-PHYSICS MANIFESTO — FOR THE FUTURE OF PHYSICS AND BIOLOGY

Y. OONO

Department of Physics and Institute for Genomic Biology,
University of Illinois at Urbana-Champaign
Urbana, Il 61801, USA
yoono@uiuc.edu

The Newtonian revolution taught us how to dissect phenomena into contingencies (e.g., initial conditions) and fundamental laws (e.g., equations of motion). Since then, 'fundamental physics' has been pursuing purer and leaner fundamental laws. Consequently, to explain real phenomena a lot of auxiliary conditions become required. Isn't it now the time to start studying 'auxiliary conditions' seriously?

The study of biological systems has a possibility of shedding light on this neglected side of phenomena in physics, because we organisms were constructed by our parents who supplied indispensable auxiliary conditions; we never self-organize. Thus, studying the systems lacking self-organizing capability (such as complex systems) may indicate new directions to physics and biology (bio-physics).

There have been attempts to construct a 'general theoretical framework' of biology, but most of them never seriously looked at the actual biological world. Every serious natural science must start with establishing a phenomenological framework. Therefore, this must be the main part of bio-physics. However, this article is addressed mainly to theoretical physicists and discusses only certain theoretical aspects (with real illustrative examples).

Keywords: Contingencies; phenomenology; complexity; Darwinism; cell theory; eucarya.

1. Introduction

It is said that this is the century of biology. Many physicists are working on problems apparently related to biology; biophysics is a fashionable branch of physics. I believe physics is a discipline not defined by what it studies, but by how it studies the world. We physicists should not be confined to the conventional interpretation of physics as the study of 'physical' world mostly excluding animated objects and their epiphenomena including the humanities. The so-called complex systems study is supposedly a discipline to take care of this 'unphysical' world. However, have complex systems study and biophysics been really useful to understand our world? This is the question behind this article. Advocating physics imperialism in its ultimate form might be a hidden agenda.

In Section 2, I will discuss what is wrong with the complex systems study practice[1] and then, in Section 3, look at what physics has not been doing. After these preliminary considerations, in Section 4, I point out the important characteristic of genuinely complex systems including organisms. This tells us where to focus our attention to build a phenomenology of biological world (Section 5). Section 6 discusses a necessary condition for complex systems, and Section 7 discusses two qualitative consequences of the phenomenological survey of the world of organisms.

2. What is wrong with Complex Systems Study?

To understand the real difference between simple systems and (genuinely) complex systems let us compare (A) a droplet of saturated salt solution with a floating small salt crystal and (B) a droplet of water containing a single cell of *Escherichia coli*. It is easy to write down the equations of motion for these systems; we have only to write down the Schrödinger equation:

$$i\hbar\frac{\partial\psi}{\partial t} = \left\{-\sum_i \frac{\hbar^2}{2m_i}\frac{\partial^2}{\partial \boldsymbol{r}_i^2} + \sum_{i>j}\frac{q_i q_j}{4\pi\epsilon_0|\boldsymbol{r}_i - \boldsymbol{r}_j|}\right\}\psi. \tag{1}$$

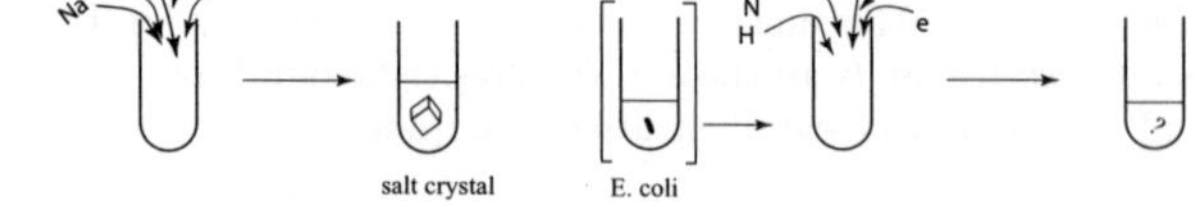

Fig. 1.1 Analogue experiments to distinguish simple and complex systems.

To this end we need only elemental analyses: we need the numbers, mass m_i and the charge q_i of various species of nuclei and the electron charge and mass. Of course, we cannot actually write the equations down, but it is clear that between (A) and (B) exists no fundamental difference thanks to the linearity of the fundamental laws. There is every reason to believe that both (A) and (B) satisfy (1). To confirm this we have 'only' to solve the equations. There is no hope of doing this digitally in the foreseeable future, but we can do quantum computation: analogue computation or actual experiments.

Can we obtain (A)? Since (1) is a partial differential equation, we need auxiliary conditions: a boundary and an initial condition. In this case, the boundary condition is not so important. Since we can expect that a cell of *E. coli* stays alive at least for a short time in a completely isolated water droplet, we may assume that the boundary conditions are homogeneous Dirichlet conditions. In case (A) we expect that 'almost all' the initial conditions with appropriate energy give a salt water droplet containing a small salt crystal. This is exactly the reason why equilibrium statistical mechanics works without any particular specification of the initial condition.

The case (B) is a futuristic version of Pasteur's famous experiment refuting the spontaneous emergence of life; we cannot do well with a generic initial condition. Since we cannot revive a mechanically destroyed *E. coli* cell, it is clear that structural (geometric) information is crucial. For example, it is well known that the bacterial cell wall cannot be constructed spontaneously. It is very unlikely that ribosome can be constructed spontaneously from its parts. Notice that to fold a protein numerous chaperones (folding catalysts) are usually required.[2]

Now, we clearly understand why all life is from life. Organisms lack self-organizing capability. We should recognize that self-organization is a telltale sign of simplicity. Something can happen spontaneously, because there are virtually not many ways to unfold the system or phenomenon. Unfortunately, however, often self-organizing property has been regarded as an important characteristic of complex systems. For example, Levine says:[3] "By self-organization I mean simply that not all the details, or "instructions" are specified in the *development* of a complex system." That is, he emphasizes that complex systems are characterized by the non-necessity of all the details to develop. Our emphasis point is fairly different. Needless to say, there are many details that are not required to be specified, but the existence of an indispensable core of (numerous) conditions that must be specified in detail is an important key feature of complex phenomena and systems.

If we ignore this distinction between (A) and (B), we will never understand the crucial nature of organisms. This point is completely ignored by Prigogine and Nicolis.[4] They emphasized that the difference between life and nonlife was not so large as had been thought. Thus, we physicists could relatively easily redirect our energy without any serious possibility of danger. However, as the readers have already sensed, this is a fundamental error that has misled complex systems study.

'Complex systems' require nontrivial auxiliary conditions. However, even if we impose a nontrivial initial condition, not all the systems can exhibit nontrivial results. For example, if we have a very large 2D Ising model, we can draw a very detailed gray-scale Mona Lisa on it (assuming, say, up spins are black pixels and down white). If the system evolves according to the Glauber dynamics, the masterpiece would disappear before very long. This indicates that systems like (B) must be able to store the initial information. That is, if a system can be called a complex system at all, it must be able to behave as a memory device. This observation is rather important, because memory is usually carried by broken symmetry. A complex system must be able to respond effectively to many initial conditions, so it must be a system with a lot of broken symmetries. Organisms must be full of such symmetry breaking processes whose outcomes must be specified carefully. This is, however, not at all a new point of view. Waddington clearly recognized this as illustrated by his epigenetic landscape (Fig. 2.1).

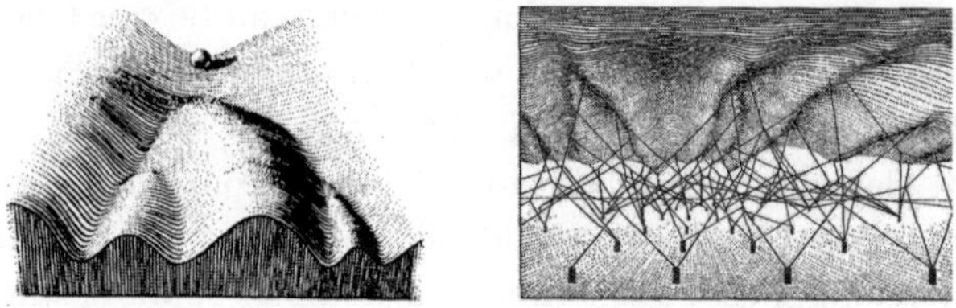

Fig. 2.1 Waddington's picture of developmental process as a cascade of symmetry breakings. Incidentally, the right depicts how genes support this landscape. Both illustrate deep ideas.[5]

The so-called emergent property is often an epiphenomenon of symmetry breaking. Prototypical examples are the position and the direction of the salt crystal in (A). It is correct that emergent properties are crucial for complex systems just as self-organization is. However, again the recognized reason is wrong. Emergent properties are important because they are indeterminate and can be specified by the initial conditions.

It should be clear that the so-called complex systems study has failed to recognize the most important prerequisite for complex systems. This failure has, however, a deep root in the history of the modern physics (after Newton).

3. What has Physics not done?

What is the structure of the Newtonian Revolution? It may be schematically written as

Phenomenon = Auxiliary Conditions + Universal Laws.

This is well-illustrated by the Newton's equation of motion for the Kepler problem:

$$m\frac{d^2\boldsymbol{r}}{dt^2} = -G\frac{M_\odot m}{r^2}\frac{\boldsymbol{r}}{r}, \tag{2}$$

where $M_\odot$ is the mass of the central star and m that of a planet. $G, t, \boldsymbol{r}$ have the usual meaning. Since this is an ordinary differential equation (ODE), we need an initial condition. In this case it is not hard to obtain accurate auxiliary conditions, so the discovery of the fundamental law (the ODE) was received with amazement. Also this resounding success seems to have determined the general direction of physics: to find fundamental laws (FL). The ODE (2) still contains contingent terms such as $M_\odot$, G, etc. Let us reduce contingent elements from FL. This is the movement illustrated in Fig. 3.1. The modern high energy physics advocates the ultimate version of this philosophy.

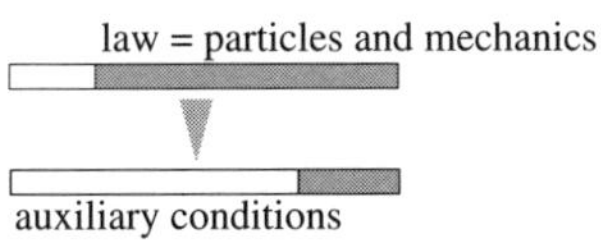

Fig. 3.1 The history of modern physics is to make the fundamental laws as pure as possible.

String theory wishes to squeeze out all the contingencies from FL. Thus, it is a popular idea that the progress of physics is in the direction shown in Fig. 3.1. What is the consequence? If we push the white-black boundary to the right, we need more 'contingencies.' Even to explain the four fundamental forces we need symmetry breaking processes. That is, to explain the particular world we live in, we must add contingencies (a lot). The growing white portion in Fig. 3.1 will be left intact by (cutting-edge) physics.

We have already noted in the last section that there are very interesting systems including ourselves that require a vast amount of auxiliary conditions. Not only the amount is vast, they are also much harder to obtain or to specify than FL as (B) exemplifies. Up to now contingencies are simply ignored because it does not seem to give universal features physicists love to find. However, there might be universal features there; at least to find universal features in the 'white' region could be an important future direction of natural science, because this region is almost everything in the future.

Even for organisms not all the auxiliary conditions required by eq.(1) need be specified uniquely. Obviously, we need not specify the position of every atom. However, we must clearly specify the auxiliary conditions that specify sectors after symmetry breaking. Let us call such indispensable auxiliary conditions Fundamental Conditions (FC). To understand a phenomenon from physicists' point of view is to understand FL and FC (see Fig. 3.2).

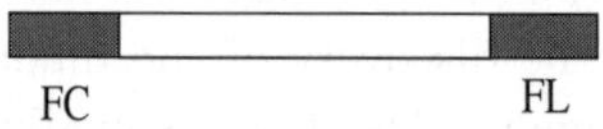

Fig. 3.2 We wish to understand Fundamental Conditions (FC) and Fundamental Laws (FL).

4. Fundamental Conditions

If we are interested in genuinely complex systems, we should concentrate our attention to FC. Therefore, in this section let us exhibit preliminary considerations on FC.

In the preceding section auxiliary conditions may be entitled to be called FC that specify sectors after symmetry breakings. In this sense, even the system (A) has a room to accommodate some FC (to specify the position of the crystal and its orientation). In this case to find out FC is not very hard. Therefore, from now on we pay due attention to FC that contains numerous conditions (i.e., ideally, we take a sort of 'thermodynamic limit"). Thus, auxiliary conditions satisfying the following two conditions are FC:[a]
(FC1) FC must be uniquely specified to realize system's characteristic features; especially they must specify the fate of the system after symmetry breaking processes.
(FC2) FC cannot emerge spontaneously (within the characteristic time of the system).
The second condition implies that history and tradition are crucial.[b] Often physicists hate history. However, we should listen to Ortega stressing the

[a]I have no intention to confine FC to be characterized by these two conditions only.
[b]To respect the complexity of our society is to respect tradition as Hayek stresses. We must not forget to pursue the consequences of (corrected) complex systems study in the humanities as well.

importance of universality.[c] If he is right, we may believe in universal features in biology.

First, we assume a principle that there is no information not carried by material structures (microscopic no-ghost principle).[d] Classification of FC is an important topic, but no systematic consideration has been given yet, so here we will be contented with a rather informal classification: If a structure itself must be specified in the auxiliary condition and if the specified structure itself function as information (say, as a template), we call the FC a structural FC (SFC). If a specified structure is used as a symbol,[e] we call it a symbolic or instructional FC (IFC). Although it must be stressed that FC for an organism is not exhausted by the genome (the totality of DNA in a cell),[f] let us take a genome to illustrate SFC and IFC. A genome consists of

C: coding part

Cs: structural — e.g., structural proteins, enzymes

Cr: regulatory — e.g., transcription factors

NC: non-coding part

I mean Cr + NC is IFC and Cs is SFC. Probably, the genome is the easiest part to handle among all the FC required by an organism as can be seen from a recent whole genome replacement experiment.[7]

Informally, an organism requires FC, but this FC cannot be produced *de novo* (cf. FC2). If constructing an organism in this world is analogized as Biology solving a problem posed by Nature, without FC Biology cannot solve the problem. Thus, we must regard the problem posed by Nature very hard. To solve it within a short time FC is required as a sort of an oracle set in the sense used in the theory of computation.

Let us tentatively characterize a complex system as a system at least

[c] "It is true that it is only possible to anticipate the general structure of the future, but that is all that we in truth understand of the past or of the present. Accordingly, if you want a good view of your own age, look at it from far off. From what distance? The answer is simple. Just far enough to prevent you seeing Cleopatra's nose." (Ortega, *La rebelión de las masas* (1930), p55). This is nothing but an expression of his belief in universality.

[d] However, there are informations carried by emergent structures; they look rather like ghosts, so the adjective 'microscopic' is attached. It may well be the case that the general chaperone atmosphere or that of the genomewide methylation condition can collectively carry important cues. In this paper this important topic will not be discussed.

[e] in the sense of C. S. Peirce

[f] Recall even irradiation damage that kills a bacterial cell is not on its DNA but on its proteins.[6]

requiring FC. Then, it is likely that
'Theorem' D. A complex system can die and never be resurrected.
'Theorem' E. If a complex system has existed for a sufficiently long time, it must be maintained by Darwinism.
'Theorem' C. A complex system that requires SFC must be spatially locally segregated.
Here, quotation marks imply that these propositions are hoped to be theorems eventually in the true sense of this word. At this level of insufficient formalization, however, they are pseudotheorems summarizing informal arguments.

We need several basic assumptions about the world:
A1. Ubiquity of Almighty Noise that damages/degrades everything including FC.
A2. Available information is always space-time local and rather small (incomplete).

A2 implies that the law of large numbers (LLN) cannot be used fully to beat noise. A1 means that FC will be damaged sooner or later. Then, A2 + FC2 implies that damaged FC cannot be repaired in the long run. Hence, 'Theorem' D.

Then, how can complex systems continue to exist despite noise? Damaged FC must be restored to the level of being capable of forming a viable organism. However, there is no correct FC posted anywhere. Since no large scale information collection is possible, LLN cannot work fully and sooner or later noise destroys the original FC. Thus, comparing several FC does not guarantee to restore the uncorrupted FC. Furthermore, the comparison of FC is actually not practical, either, as can been seen from the following consideration. We must recognize that the analogy between making an organism and solving a hard question posed by Nature is rather deep. As is well known, hard problems are represented by NP complete problems. For such problems finding a solution is hard, but to check the correctness of a solution is easy. In our context to check that FC gives the right solution is equivalent to forming a viable organism. Of course, we know developmental process is 'easy' (far easier than the evolution process that created FC). Thus, comparing FC without forming actual organisms is computationally inefficient.

Suppose a corrupted FC is known to fail to give a viable organism. How can an uncorrupted FC be restored? Correcting corrupted solution to a hard problem is again a hard problem.[g] Therefore, given a corrupted FC, whether 'supposedly repaired versions' are really repaired or not can be checked only through using them to form organisms. Notice that this procedure is non-deterministic in the sense of the theory of computation. We must produce (supposedly) repaired FC randomly (or with a stochastic process whose Kolmogorov complexity is at least comparable to that of the noise in the world; as we know the actual biological systems use the *same* noise that can destroy them). Then, we 'run' them to form corresponding viable organisms. How can we find the corrected outcome? By comparison, but comparison with what?

We need another assumption as to the finiteness of the world, or as to the finiteness of the available resources:
A3. We cannot sustain indefinitely many organisms.
Inevitably, comparison becomes competition and entails selection.

Our argument up to this point may be summarized as follows. If we assume that complex systems exist for sufficiently long time, then FC is maintained against noise by nondeterministic computation. This implies reproduction and mutation are absolutely needed to *maintain* complex systems for a long time. Because of the bounded resources this process entails selection. Thus, if there has been a complex system for a sufficiently long time, it must have been sustained by Darwinism. This is 'Theorem' E. However, Darwinism does not imply the presence of complex systems as was demonstrated by Spiegelman's monster.[8]

FC must contain SFC, if we consider a complex system that arises from the microscopic molecular scale. On this basis can symbolic or IFC work.
(1) SFC provides templates/nuclei to guide self-organizable materials. If repetitive structures are barred and if there is no IFC, the size of the resultant system is comparable to the size of SFC.
(2) Structure itself is the information for SFC; the components of SFC must keep well-defined spatial relations, so SFC must be localized in space.
(3) The SFC for a system and SFC for other systems must be clearly distinguished.
Therefore, complex systems requiring SFC must be locally segregated in space. If it is not tethered (e.g., as a single polymer), it must be enclosed in a well-defined domain. In this sense cellularity is required by complex systems. This is 'Theorem' C.

According to Barbieri[9] the two pillars of the modern biology are Dar-

[g]Notice that there is no incremental (or recursive) way to solve NP hard problems.

winism and Cell theory. Notice that complexity implies both. There are two versions of cell theory, weak and strong. The weak version asserts that the organisms we know are made of cells. Here, 'Theorem C' asserts the strong version: all organisms are made of cells.

5. How to obtain Phenomenology of Complex Systems

We might be able to push such a deductive study further, but our imagination and logical power are very limited. Every empirical science must start with phenomenology. Here, I mean by this word a summary of what we observe free of any ontological bias. We should clearly recall that without thermodynamics, equilibrium statistical mechanics could not be created/cannot be formulated; because there is no well defined phenomenology nonequilibrium statistical mechanics does not exist. If we wish to understand a class of phenomena or systems in general terms (in a non-fetish fashion, so to speak), we need good phenomenology. To have a phenomenological summary we must collect facts. Therefore, I must start with surveying the world of organisms. Most physicists may call this stamp collecting, following Rutherford, but the hope here is that even stamps may tell us important lessons if sufficiently accumulated. However, already G. G. Simpson warned in *The Meaning of Evolution*: "Indeed facts are elusive and you usually have to know what you are looking for before you can find one." Thus, my strategy is to organize observed facts around FC. This project does not aim at contributing biology but primarily at enriching physics (and perhaps mathematics), but as a physics imperialist I firmly believe that this is the only way to understand biology properly.

Fundamental questions about FC include the following:

(1) How is FC organized? e.g., organization of genome.

(2) How is FC used?
development, aging, — at the organismal time scale.
behavior, defense, homeostasis — at the time scale of sec to min.

(3) How has FC been changing?
History—paleontology and evolution.
Resultant diversity—taxonomy.
Collective phenomena— ecology, sociobiology.

We deliberately ignore the aspects directly connected to materials. This implies to ignore a large portion of biophysics as irrelevant. We have seen that complex systems require a lot of symmetry breaking processes that are prepared by self-organizing properties of materials. Therefore, naturally there are two research directions: (i) emphasizing self-organizing capability of materials, (ii) emphasizing indeterminate aspects produced by self-organization. Biophysics stresses (i); it is largely molecular and materials science of biological matter; dead bodies, albeit fresh, are enough; physics of minced meat. In contrast, I stress (ii), because how to utilize the emergent indeterminacy is the key to complex systems.

The inductive part of phenomenology consists of two parts: (PI1) Compiling facts and (PI2) Distilling phenomenology from the facts. The main part of PI1 is, for physicists, to develop new (methods and devices to aid) experiments and field work. If you are interested in biology seriously, you should have a taxonomic group you are familiar with. Natural history is very important, because we are interested in universality. We can find universality only through comparative studies. In this article I do not discuss any experimental aspects, but I wish to emphasize the importance of developing high-throughput phenomic studies (in contrast to the genomic studies already in bloom). We are interested in organisms, not in molecules *per se*; molecules make sense only in the light of natural history. This is why I must stress phenomics. PI2 is the data analysis, text-mining, etc. This is also crucial because we are inundated with numbers from high-throughput experiments.

We should look for exact phenomenologies like thermodynamics that allows us to make precise predictions. I do not have such a well defined phenomenology yet. However, some general observations I have may already be of some use. As an example, in the next section, I outline presumably the most common complexification process.

6. Basic Observations about FC and Complexification

What is complexity? Our provisional necessary condition for a complex system is that it requires FC for its construction. Therefore, a certain quantitative measure of FC might characterize the complexity of a system. However, probably it is the consensus that complexity has many facets, so organisms are not well ordered with respect to complexity.[10] Therefore, it may be expected that no single important complexity measure exists. However, Fig. 5.1 is an interesting observation about the non-coding DNA.[11]

A remarkable message of Fig. 5.1 is that the usual anthropocentric view point detested by Gould[12] (and Woese[13]) seems vindicated. From the point of view of FC, the amount of IFC is a good measure of complexity. This view point is consistent with the evolution of micro RNA.[14] A natural logical consequence is, as can be seen from Fig. 5.1, that study of complexity

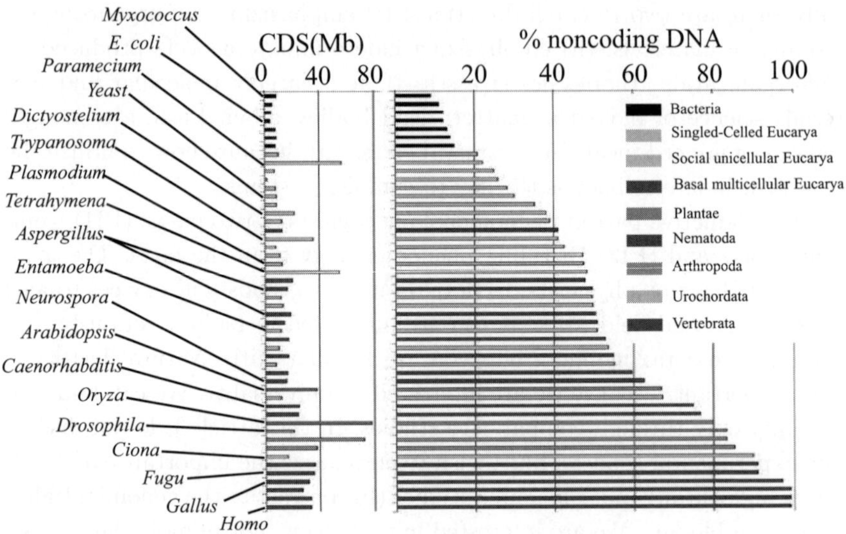

Fig. 5.1 If we pay attention to the amount of non-coding DNA in the genome, organisms are ordered naturally in the usual 'anthropocentric' order which Gould detested. CDS is the amount of protein coding DNA in megabase. The figure is due to Taft et al.[11]

or biocomplexity must be the study of Eucarya.

Complex systems may be classified into two major classes; one that mainly relies on SFC alone, and the rest that relies on IFC as well. The Procarya/Eucarya dichotomy roughly corresponds to this distinction. Thus, even though Procarya is a paraphyletic group, it may be mathematically a well defined natural group. It should be recognized that spontaneous formation of a 'large' system is impossible with Brownian motion + SFC alone.[h] If a large system requires FC, it requires IFC. This also implies that Procarya is not really interesting from the complexity point of view.[i]

[h]One might say dissipative structure may evade such constraints. However, dissipative structures without microscopic materials organization change are too fragile to be relevant to biology. Those with materials bases are essentially equilibrium structures modulated by dissipation. Thus, dissipative structures are basically irrelevant to biology.

[i]One might say that from the biodiversity point of view Procarya is crucial. We could say where there is a free energy difference there is a prokaryote exploiting it. However, this is a diversity of organic chemistry; if we change methyl to ethyl to propyl to $\cdots$, we could make a diverse set of reactions and compounds. Thus, I bet that only in this sense Procarya is diverse, so from physicists' point of view a simple universal picture might be obtainable for the whole Procarya.

We know FC can be made only through Darwinism.[j] Then, how does it evolve? A possible question is: Can FC grow gradually? I do not know any such example. Requiring new symmetry breakings may be at odd with continuity. My conclusion is summarized in Fig. 5.2.

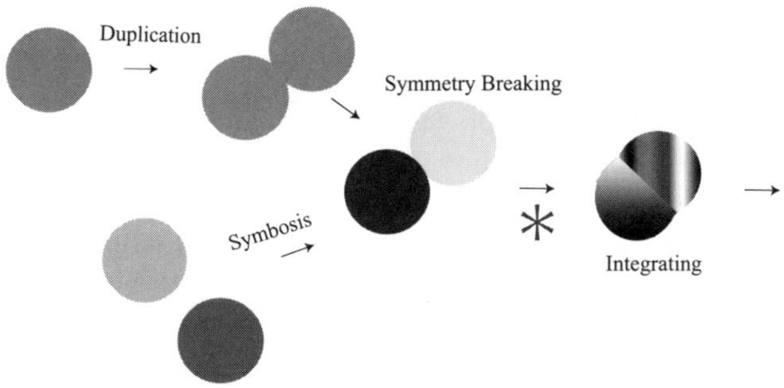

Fig. 5.2 A unit process for complexification: it consists of two steps; the duplication/juxtaposition step before * and the much more important integrating step *.

Complexification occurs in two steps. Complexification requires symmetry breaking (or rather numerous symmetry breaking processes). Symmetry breaking requires a stage that can accommodate various sectors created by symmetry breaking processes. Expanding the stage is often due to symbiosis (juxtaposition) or due to duplication/multiplication. The importance of duplication and subsequent subfunctionalization/neofunctionalization was stressed by S. Ohno.[15] The importance of symbiosis has been stressed by Levine.[16] Major historical events other than mass extinction may have been driven by establishment of new symbiosis. For example, the landfall of green plants must have been due to plant-fungi alliance. There are numerous such examples.

This is the first of the two major steps making a unit process of complexification. The second step is integration. There is no good name for this step, because the real importance of this step has not fully been recognized as a general process. In short, what happens is to use all the elements created by the first step in order to realize a higher level organization. Even in the case of endosymbiosis, the process of incorporating endosymbionts into

[j]Even if God were to exist and to have created organisms, to maintain them against Almighty Noise we need Darwinism. Consequently, even the initial intelligent design is meaningless under Almighty Noise.

the host cell as new cellular organelles to create a new type of cells may be understood as an example of this second step. Incidentally, most Procarya cannot afford duplication due to its sheer size. This is also a reason why there is not much complexification in Procarya; the most important complexification path is blocked.[k]

The second step of the complexification process is the crucial step. The first step is often a preparatory step. This step can quantitatively increase parts and functions, but qualitative changes may not occur there. The idea is supported by the formation of, e.g., Metazoa and Bilateria. We now know that Choanoflagellata has[17] many signal pathway components and cellular communication molecules that are organized and utilized by Porifera.[18] Even Anthozoa (Cnidaria) has (and probably Porifera had) Hox genes;[19] Hox genes are used to make the bilaterian body plan. Another example is our language. It is highly likely that all the components required by the linguistic capability exist in primates. Therefore, the rate process for the emergence of language could have been the integration step. The lesson is: some sort of 'nucleation process' that starts to integrate preexisting key components is really the crucial step to achieve higher level complexity. Even the evolution of society and civilization could be understood along this line. This is the step marked with $*$ in Fig. 5.2.

It is often said that excessively specialized organisms cannot evolve. Perhaps, our general consideration sheds some light on this folklore. The complexification process consists of two steps. If an organism loses many elements created in the first step, the integration step would be virtually aborted or at best incomplete. In this sense, complexification occurs most likely in the lineage preserving most primitive (or plesiomorphic) features. Loss of features prepared during the first step seems to be the key ingredient of 'specialization.' Furthermore, many examples tell us that an efficient way to lose these features is the sessile and/or filter feeding life style (or the loss of capability to move around[20]). The observation is supported by our position in Deuterostomia.[l] Echinodermata and Hemichordata are spe-

---

[k]One might say that extending the biofilm and other multicellular structures even Procarya could complexify. However, this is highly unlikely due to frequent adaptive sweeps.

[l]For convenience, some classification rudiments are given here. We vertebrates are in Chordata containing Cephalochordata and Urochordata (sea squirt, etc.) as well. Chordata is among Deuterostomia with Xenoturbellida, Echinodermata (sea urchins, sea stars, etc) and Hemichordata. Deuterostomia is among Bilateria (including most of invertebrates). Bilateria and Cnidaria (sea anemone, hydra, etc.) make up Eumetazoa, which with Porifera (sponge) makes the major portion of Metazoa (= Animalia). The closest sister group in Opisthokonta to Metazoa is Choanoflagellata. Opisthokonta includes

cialized branches compared with Chordata. Within Chordata notice that Cephalochordata that can move around is the most primitive to which we (Vertebrata) are close; Urochordata are sessile filter feeders, so they are specialized. Thus, we humans are in the lineage of the least specialized within Deuterostomia. The same may be said about Chordata among Metazoa. We can expect that actively moving creatures were the common ancestors of Calcarea and Eumetazoa, so we came from something like planulae. Porifera are sessile filter feeders, a dead end from the complexification point of view. Notice that the Planulozoa-Porifera relation reminds us of the Cephalochordata-Urochordata relation. Thus, we humans are in the lineage of the least specialized within Animalia. The recent *Nematostella* genome[21] supports this point of view. Where is then Opisthokonta that includes Animalia within Eucarya? It is likely that Unikonta is the basic group. Again, we are in the group basic to Eucarya.

To simplify, we may say that the first expanding stage of the complexification process prepares a (wide) stage and actors. The second step gives scenarios. IFC is crucial in this step. Specialization implies loss of actors (and a shrinking stage) before any interesting play begins. Sessile life style is an efficient way to decimate actors.

7. Potential use of Qualitative Phenomenology

There have been attempts to construct a 'general theoretical framework' of biology, but most of them never seriously looked at the actual biological world. Thus, these studies are not so interesting to biologists.

The program I am proposing may be called Integrative Natural History that unifies molecular, phenomic and much larger scale observations to understand genuine complex systems in an unified fashion.[m] Its theoretical (deductive) side consists of two parts: (PD1) Constructing phenomenological theory of complex systems as mathematics and (PD2) Formulating many biologically meaningful questions based on the phenomenological summary.

Although I do not have any precise phenomenology, it seems possible to say something on the PD2 side. For example, we have already seen that major historical events other than mass extinctions are likely to be driven

Fungi and us and is one of a few kingdoms of Eucarya. In Domain Eucarya Opisthokonta is among Unikonta with Amoebae.

[m]This unification has a much more significant implication in biology, because oragnisms are in a certain sense inflated microscopic systems, quite different from many systems physicists have been studying that have layered structures with separated micro, meso and macroscopic levels.

by symbiotic relations that open up new (vast) niches. Then, a natural big question is whether there was a major symbiotic relation driving the landfall of animals. Gut symbionts must be paid due attention.

We have also seen that organisms with most plesiomorphic traits complexify most; we are directly related to the most basic group of Eucarya. What is the natural 'analytic continuation' of this idea? The most natural conclusion is that (anaerobic) Eucarya is the basic organism, and Procarya is a specialized group that lost many plesiomorphic features. This idea may sound crazy, but notice that the common notion that Procarya (Bacteria) was earlier than Eucarya has no unambiguous supporting fact. This crazy idea may well be consistent with the evolution of codons[22] and introns.[23]

I wish to conclude the article with two more such general observations.
(1) Biological systems lack foresight.
This may be called the 'principle of non-determinism' (in the theory of computation sense). This is well understood and not at all a new observation. Evolution process itself is a great example. Overproduction combined with subsequent thinning is a common strategy in neural development. There is every reason to believe that molecular machines work on this principle: the right outcome is not aimed at. Instead, when the desired outcome is realized, it is stabilized (preserved). Even the translation of a DNA sequence to the corresponding amino acid sequence relies on this strategy. Thus, we may say, molecular machines work on the principle of motional Darwinism. This implies that driving with some potential is not essential for molecular motors. Only some steric hindrance forbidding some class of movements is needed.
(2) Biological systems cut Gordian knot(s), or biology never solves difficult problems (the Gordian Principle?). There are many mathematically difficult problems apparently relevant to the biological world.

A famous example is the protein folding problem: to determine the tertiary structure from the primary sequence. Since there is an astronomical number of conformations possible for a given primary sequence, Levinthal pointed out that folding should be combinatorially very hard. However, we know many proteins are formed very quickly *in vivo*. Has biology solved this problem? No. For short chains, Biology picked up special quickly foldable sequences, and then connected them to make longer chains. Thus, proteins consisting of up to 200-300 amino acids are based on special quickly foldable proteins. The problem is not solved generally at all. For longer chains Biology has given up, and relies on chaperones (folding catalysts).

Another famous problem is the DNA entanglement, a topological difficulty. When prokaryote DNAs were found to be ring-shaped, mathematicians were delighted, expecting topological invariants to interfere replications, etc. However, such a topological difficulty is non-existent thanks to, especially, topoisomerase II.

Statistical physicists discussed vesicle shapes, expecting such problems were biologically relevant (think of ER): how can complicated shapes be formed 'spontaneously'? However, no differential geometrical problem of manifolds arises, because these membrane shapes are in detail controlled even stoichiometrically by membrane associated proteins, e.g., clathrin, caveolin, coatomer complex, etc.

These may be rather disappointing stories for mathematically oriented theoreticians, but we should learn an important lesson. Whenever we can expect difficult problems in biological processes (or rather, processes relevant to complex systems), the real problem for theoreticians is to think how Biology avoids solving or even facing them. I wish to point out three such examples. I believe all are important problems.

(i) We can expect combinatorial difficulties in regulation of genes. Therefore, Biology must have worked hard to avoid unnecessary combinatorial entanglements. For example, after duplication, gene-gene relations that may cause entanglements are selectively suppressed.[24]

(ii) We can make extremely highly nested clause structure 'formally,' so this recursiveness has been stressed as an important feature of our natural language. However, as is well known, self-referencing can cause numerous paradoxes such as the liar paradox. How do our brains avoid such logical difficulties? We should reflect on our daily language practice; we usually do not decide whether we decide or not whether we should go to a meeting or not, for example. Thus, it is highly questionable that our brain honestly handles recursiveness as such.

(iii) If you read contemporary ethics textbooks, you will certainly find many serious ethical aporias. However, if we take seriously the lesson Biology teaches us, we ought to realize that the most important problem of ethics is to avoid the situations of ethical aporias; we should work hard to avoid hard problems! Although the current environmental problems are not simply due to the population problem,[25] still there is no doubt that the human population problem will be the most important factor that will destabilize our world. How do our fellow creatures cope with this problem? It is known that the actual wild population is far less than the environmental capacity (self-limitation is observed[26]).

Sociopolitical issues should not be avoided by physicists.

References

1. Y. Oono, Int. J. Mod. Phys. B, **12**, 245 (1998).
2. See, for example, J. C. Young, V. R. Agashe, K. Siegers and F. U. Hartl Nature Reviews Molecular Cell Biology **5**, 781 (2004).
3. S. A. Levine, *Fragile Dominion* (Perseus Publishing, New York, NY, 1999).
4. G. Nicolis and I. Prigogine, *Exploring Complexity: an introduction*, (W. H. Freeman, New York, 1989).
5. C. H. Waddington, *The Strategy of the Gene* (Allen and Unwin, London, 1957).
6. M. J. Daly, E. K. Gaidamakova, V. Y. Matrosova, A. Vasilenko, M. Zhai, R. D. Leapman, B. Lai, B. Ravel, S.-M. W. Li, K. M. Kemner, J. K. Fredrickson, PLoS Biol., **5**, e92 (2007).
7. C. Lartigue, J. I. Glass, N. Alperovich, R. Pieper, P. P. Parmar, C. A. Hutchison III, H. O. Smith, J. C. Venter, Science, **317**, 632 (2007).
8. D. L. Kacian, D. R. Mills, F. R. Kramer, and S. Spiegelman, Proc. Natl. Acad. Sci., **69**, 3038 (1972).
9. M. Barbieri, *The Organic Code, An introduction to semantic biology* (Cambridge, 2002).
10. D. W. McShea, Evolution **50**, 477 (1996).
11. R. J. Taft, M. Pheasant, and J. S. Mattick, BioEssays **29**, 288-299 (2007).
12. S. J. Gould, *The Structure of Evolution Theory* (Harvard UP, 2003), esp., p897 and p1321.
13. C. R. Woese, Proc. Nat. Acad. Sci., **95**, 11043 (1998).
14. R. Niwa and F. J. Slack, Curr. Op. Gen. Dev., **17**, 145 (2007); L. F. Sempere, C. N. Cole, M. A. McPeek, and K. J. Peterson, J. Exper. Zool., B **306**, 575 (2007).
15. S. Ohno, *Evolution by Gene Duplication* (Springer, New York, 1970).
16. S. A. Levine, PLoS Biol., **4**, e300 (2006).
17. N. King and S. B. Carroll, Proc. Natl. Acad. Sci., **98**, 15032 (2001); N. King, C. T. Hittinger, and S. B. Carroll, Science **301**, 361 (2003).
18. W. Dirks, J. S. Pearse, and N. King, Proc. Nat. Acad. Sci., **103**, 12451 (2006).
19. K. J. Peterson and E. A. Sperling, Evol. Dev., **9**, 405 (2007).
20. The importance of movement in the compleification process has been stressed by T. Ikegami. See, for example, his recent book, *Movement Creates Life* (Seido-sha, 2007) [In Japanese].
21. N. H. Putnam, M. Srivastava, U. Hellsten, B. Dirks, J. Chapman, A. Salamov, A. Terry, H. Shapiro, E. Lindquist, V. V. Kapitonov, J. Jurka, G. Genikhovich, I. V. Grigoriev, S. M. Lucas R. E. Steele, J. R. Finnerty, U. Technau, M. Q. Martindale,7 D. S. Rokhsar, Science, **317**, 86 (2007).
22. S. Itzkovitz and U. Alon, Genome Res., **17**, 405 (2007).
23. A. M. Poole, D. C. Jeffares and D. Penny, J. Mol. Evol., **46**, 1 (1998).
24. G. C. Conant and K. H. Wolfe, PLoS Biol., **4**, e109 (2006).
25. J. Vandermeer and I. Perfecto, *Breakfast of Biodiversity, the political ecology of rain forest destruction*, 2nd Edition (Food First Books, 2005).
26. J. E. C. Flux, Oikos **92**, 555 (2001).

SINGLE MOLECULE FORCE MEASUREMENT FOR PROTEIN SYNTHESIS ON THE RIBOSOME

SOTARO UEMURA

Graduate School of Pharmaceutical Sciences, University of Tokyo, 7-3-1, Hongo, Bunkyo-ku, Tokyo, 113-0033, Japan

Precursory Research for Embryonic Science and Technology (PRESTO), Japan Science and Technology Agency

The ribosome is a molecular machine that translates the genetic code described on the messenger RNA (mRNA) into an amino acid sequence through repetitive cycles of transfer RNA (tRNA) selection, peptide bond formation and translocation. Although the detailed interactions between the translation components have been revealed by extensive structural and biochemical studies, it is not known how the precise regulation of macromolecular movements required at each stage of translation is achieved. Here we demonstrate an optical tweezer assay to measure the rupture force between a single ribosome complex and mRNA. The rupture force was compared between ribosome complexes assembled on an mRNA with and without a strong Shine-Dalgarno (SD) sequence. The removal of the SD sequence significantly reduced the rupture force, indicating that the SD interactions contribute significantly to the stability of the ribosomal complex on the mRNA in a pre-peptidyl transfer state. In contrast, the post-peptidyl transfer state weakened the rupture force as compared to the complex in a pre-peptidyl transfer state and it was the same for both the SD-containing and SD-deficient mRNAs. The results suggest that formation of the first peptide bond destabilizes the SD interaction, resulting in the weakening of the force with which the ribosome grips an mRNA. This might be an important requirement to facilitate movement of the ribosome along mRNA during the first translocation step. In this article, we discuss about the above new results including the introduction of the ribosome translation mechanism and the optical tweezer method.

1. Ribosome Introduction

1.1. *Translation*

Translation is the cellular mechanism of protein synthesis. The principle component in this biochemical process is a macromolecular enzyme called the ribosome. Protein synthesis, a central step in gene expression, is highly regulated and a conserved process. The mechanism of ribosome function is critical to understanding basic aspects of gene regulation at the level of translation. The life cycles of Human Immunodeficiency Virus (HIV) and Hepatitus C virus (HCV) infections depend on translational regulation events. Today, much is known about HIV and HCV mediated translation control, however, very little is understood of how these events may be effectively disrupted.

The ribosome is an unusual enzyme in that it is composed of both RNA and protein. It is a two-subunit enzyme; a large and a small subunit component join to form a functional ribosome particle. The ribosome of *E. coli* is an approximately 2.5 million Dalton, two-subunit RNA-protein complex, which is composed of three stable RNA species: 23S, 16S and 5S ribosomal RNA (rRNA) so named by their apparent sedimentation coefficients in Svedberg units. 23S and 5S rRNA molecules assemble cooperatively with 33 proteins to form the larger of the two subunits of the ribosome, sedimenting at 50 Svedberg (50S). Similarly, 16S rRNA assembles with the cooperation of 21 proteins to form the smaller subunit, sedimenting at 30S. 30S and 50S subunits join to form a functionally competent 70S particle. Assembled ribosome particles are structurally designed to interact, with messenger RNA (mRNA) and transfer RNA (tRNA) ligands. These components form the core of the translation apparatus.[1–4]

The basic function of the 70S ribosome in translation is to read genetically encoded information in the form of mRNA nucleotides and convert it into amino acid sequence. Transfer RNA covalently linked to an amino acid is an L-shaped substrate of the ribosome that functions as an adapter molecule in this process. tRNA molecules bind to the ribosome at the interface between the two joined subunits to interact with mRNA and localize their aminoacyl moieties in adjacent sites within the ribosome (Figure 1). At least three tRNA molecules can occupy the intersubunit space; binding occurs at adjacent sites called the aminoacyl (A), peptidyl (P) and exit (E) sites. All three tRNA binding sites are formed by the cooperative union of large and small subunit elements.

In this configuration, genetic information encoded within mRNA can make specific base pairing interactions with tRNA; the amino acid moieties of tRNA molecules bound at adjacent codons come into close proximity for peptide-bond formation. tRNA binding to the ribosome occurs through one of several processes collectively known as decoding; amino acid polymerization is carried out by the ribosomes intrinsic peptidyltransferase activity.

These ribosome functions are linked to the cooperative interaction of tRNA with both subunits. tRNA molecules enter the ribosome in a specific order; in general, tRNA enter at the A site and move to the P and E site in subsequent steps which is called the elongation process.

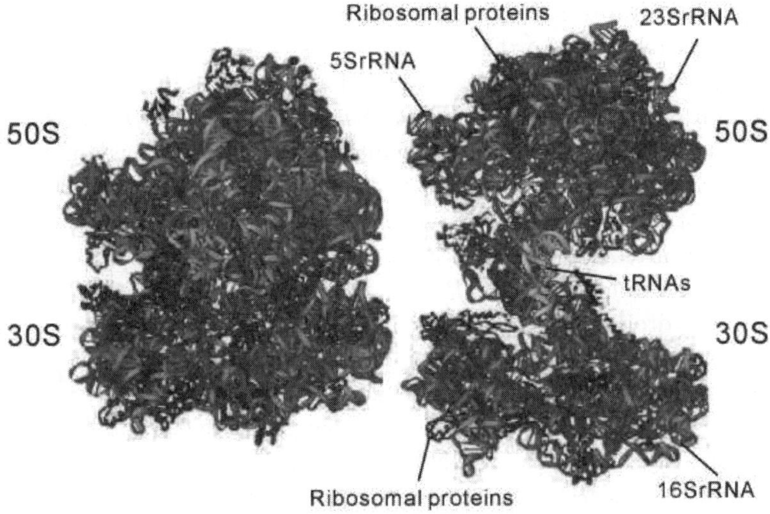

Fig. 1. 70S ribosome crystal structure. The structure of the complete Thermus thermophilus 70S ribosome was crystallized in a complex including the 30S subunit (16S rRNA and small subunit proteins), 50S subunit (23S rRNA, 5S rRNA, and large subunit proteins), P- and E-site tRNA, and solved by x-ray crystallography to a resolution of 5.5 Å (figure taken from M.M. Yusupov et al[1] 2001).

1.2. *Elongation cycle*

The elongation phase of translation is the process in which codons are paired with the proper tRNA, polypeptide bonds are formed and translocation to the next codon occurs. All of the processes of elongation are ribosome-catalyzed and can take place in the absence of translation factors. Nonetheless, elongation is driven rapidly with high fidelity by elongation factors, EF-Tu and EF-G. Contemporary understanding of elongation has been revealed through careful biochemical and structural analysis.[1–4] These data have identified the components of the translation apparatus that are important to tRNA binding and movement. Both EF-Tu and EF-G proteins having GTPase activity have their overlapping binding sites on the ribosome.

The first step of elongation is EF-Tu delivery of aminoacyl-tRNA to the A site of the ribosome (Figure 2, left). EF-Tu mediates this event by interacting with the aminoacylated acceptor stem of tRNA and the large subunit of the ribosome. With the exception of $tRNA^{fMet}$, which is the initial tRNA, all tRNA are delivered to the ribosome by EF-Tu at the A site. EF-Tu, in ternary complex with aminoacyl-tRNA and GTP, binds rapidly and reversibly to the ribosome.

Ternary complex binding at the GTPase center is followed by codon-anticodon recognition. Cognate mRNA-tRNA interactions are specifically recognized by the 30S subunit. Recognition is mediated through interaction of conserved nucleotides of 16S RNA (A1492 and A1493) with minor groove elements of three base pair mini-helix formed by proper interaction of the tRNA and mRNA.[5] Codon recognition triggers events that ultimately activate EF-Tu to hydrolyze GTP.

GTP hydrolysis is most likely concomitant with conformational rearrangement of large subunit elements interacting with EF-Tu. EF-Tu-GDP has a very low affinity for both aminoacyl-tRNA and the ribosome. Following GTP hydrolysis, tRNA is released from the ternary complex to allow the 3' end of the aminoacyl-tRNA to bind within the peptidyltransferase center. Interestingly, the slow step in the decoding process takes place following tRNA release from the ternary complex.[6] This step is called accommodation (Figure 2, top). Accommodation of cognate tRNA is $\sim$1000 times more efficient than for non-cognate tRNA; the proper selection is mediated by a $\sim$20-fold increase in the off rate of non-cognate tRNA and a 60-fold increase in the rate of cognate tRNA accommodation.[6] Both effects proofreading out wrong tRNA and induced-fit capture of the right tRNA, built the fidelity of translation on the weak specificity of the codon-anticodon interaction.

Following aminoacyl-tRNA binding at the A site, peptide-bond formation rapidly takes place. This reaction is catalyzed by components of the large subunit of the ribosome.[7,8] Peptidyltransferase activity of the 50S subunit mediates nucleophilic attack of the aminoacyl-tRNA α-amino group on the electrophilic carbonyl group linking the P-site tRNA to its amino acid.

Following peptide bond formation, EF-G-GTP replaces EF-Tu-GDP at the GTPase center of the ribosome to mediate for an additional round of elongation. EF-G causes translocation of both A and P site tRNA codon-anticodon complexes with respect to the ribosome to move the downstream codon into a vacant A site (Figure 2, right). Recent time-resolved experiments have shown that GTP hydrolysis precedes translocation and strongly accelerates the reaction,[9] supporting the notion that the active form of EF-G is actually the GDP-bound one and EF-G acts like a motor protein. This model is in strong conflict with the switch model of a G protein and has received some criticism.[10]

Posttranslocational state with tRNAs occupying P and E sites. EF-G has been released, after GTP hydrolysis in GDP form, to vacate the overlapping binding site for the next ternary complex (Figure 2, bottom).

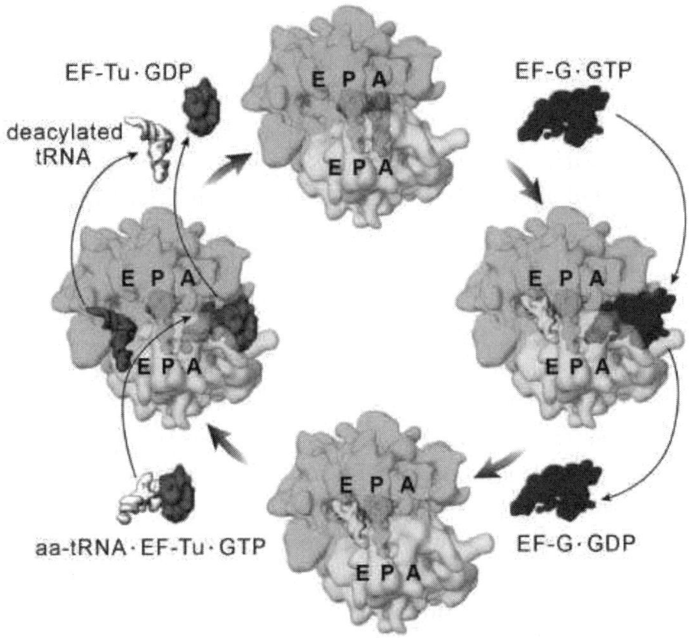

Fig. 2. The elongation cycle, showing three-dimensional positions of tRNAs and elongation factors, as obtained by cryoelectron microscopy technique, overlaid on the 1.5-nm resolution map of the Escherichia coli 70S ribosome (figure taken from M. Valle et al 2002[11]).

2. Principle for Optical Tweezers Technique

An optical tweezers is formed by tightly focusing a laser beam with an objective lens of high numerical aperture (NA). A dielectric particle near the focus will experience a focus due to the transfer of momentum from the scattering of incident photons. The resulting optical force has been decomposed into two components: (1) a scattering force, in the direction of light propagation and (2) a gradient force, in the direction of the spatial light gradient. The scattering component can be thought of pushing the bead in the direction of light propagation. Incident light impinges on the particle from one direction, but is scattered in a variety of direction, while some of the incident light may be absorbed. As a result, there is a net momentum transfer to the particle from the incident photons.[12]

For stable trapping in all three dimensions, the axial gradient component of the force pulling the particle towards the focal region must exceed the scattering component of the force pushing it away from that region. This condition necessitates a very steep gradient in the light, produced by sharply focusing the trapping laser beam to a diffraction-limited spot using an objective of high NA. As a result of this balance between the gradient force and the scattering force, the axial equilibrium position of a trapped particle is located slightly beyond (i.e., down-beam from) the focal point. For small displacements ($\sim$ 150 nm), the gradient restoring force is simply proportional to the offset from the equilibrium position, i.e., the optical trap acts as Hookean spring whose characteristic stiffness is proportional to the light intensity.

In developing a theoretical treatment of optical tweezers, there are two limiting cases for which the force on a sphere (bead) can be readily calculated. When the trapped bead is much larger than the wavelength of the trapping laser, the conditions for Mie scattering are satisfied, and optical forces can be computed from simple ray optics (Figure. 3). Refraction of the incident light by the bead corresponds to a change in the momentum carried by the light. By Newton's third law, an equal and opposite momentum change is imparted to the bead. The force on the bead, given by the rate of momentum change, is proportional to the light intensity. When the refraction index of the particle is greater than that of the surrounding medium, the optical force arising from refraction is in the direction of the intensity gradient. Conversely, for an index lower than that of the medium, the force is in the opposite direction of the intensity gradient. The scattering component of the force arises from both the absorption and specular reflection by the trapped object. In the case of a uniform bead, optical forces can be directly calculated in the ray-optics regime.[13]

When the trapped bead is much smaller than the wavelength of the trapping laser, the conditions for Raleigh scattering are satisfied and optical forces can be calculated by treating the particle as a point dipole. In this approximation, the scattering and gradient force components are readily separated. The scattering force is due to absorption and reradiation of light by the dipole. For a bead of radius a, this force is

$$F_{scatt} = \frac{I_0 \sigma n_m}{c} \tag{1}$$

$$\sigma = \frac{128\pi^5 a^6}{3\lambda^4}\left(\frac{m^2-1}{m^2+2}\right)^2 \tag{2}$$

Where I_0 is the intensity of the incident light, σ is the scattering cross section of the bead, n_m is the index of refraction of the medium, c is the speed of light in vacuum, m is the ratio of the index of refraction of the particle to the index of the medium (n_p/n_m), and λ is the wavelength of the trapping laser. The scattering force is in the direction of propagation of the incident light and is proportional the intensity. The time-averaged gradient force arises from the interaction of the induced dipole with the inhomogeneous field

$$F_{grad} = \frac{2\pi\alpha}{cn_m^2}\nabla I_0 \tag{3}$$

Where

$$\alpha = n_m^2 a^3\left(\frac{m^2-1}{m^2+2}\right) \tag{4}$$

is the polarizability of the bead. The gradient force is proportional to the intensity gradient, and points up the gradient when m>1.

Unfortunately, the majority of objects that are useful or interesting to trap, in practice, tend to fall into this intermediate size range (0.1-10λ). As a practical matter, it can be difficult to work with objects smaller than can be readily observed by video microscopy (~0.1 μm).

3. Single Molecule Force Measurement for Ribosome Complex

3.1. *Rupture force measurement method*

Structural and biochemical data for ribosome translation mechanism have revealed a network of contacts between tRNAs bound to adjacent codons in the peptidyl- and aminoacyl-tRNA sites (P and A sites) and both the 30S and 50S subunits.[1–4] Ribosome-tRNA-mRNA interactions are required for the maintenance and regulation of the ribosomal complex stability during all stages of translation, yet processive translocation requires relaxation of the interactions for movement of tRNA-codon complexes with respect to the ribosome. Initial positioning of the ribosome on mRNA involves the recognition of a purine-rich sequence, known as the Shine-Dalgarno (SD)

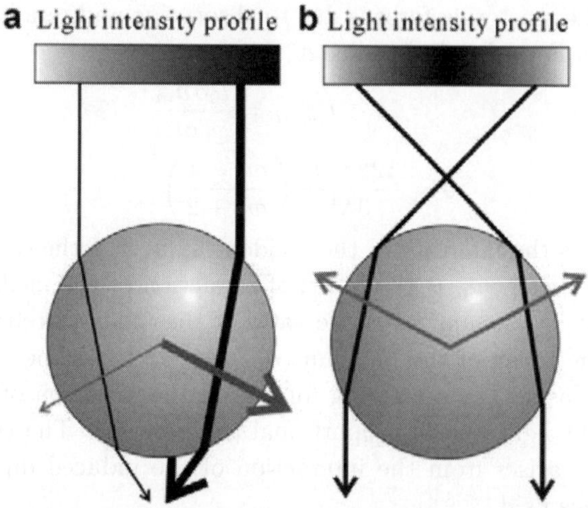

Fig. 3. Ray optics description of the gradient force. (A) A transparent bead is illuminated by a parallel beam of light with an intensity gradient increasing from left to right. Two representative rays of light of different intensities (represented by black lines of different thickness) from the beam are shown. Conservation of momentum dictates that the momentum of the bead changes by an equal but opposite amount, which results in the forces depicted by gray arrows. (B) To form a stable trap, the light must be focused. In this case, the bead is illuminated by a focused beam of light with a radial intensity gradient. Gray arrows represent the forces. If the bead moves in the focused beam, the imbalance of optical forces will draw it back to the equilibrium position.[12]

sequence, located upstream of the AUG initiation codon on the mRNA and complementary to the 3' end of the 16S rRNA.[14,15]

Although the detailed interactions between the translation components have been revealed by extensive structural and biochemical studies, it is not known how the precise regulation of macromolecular movements required at each stage of translation is achieved. It remains unclear what signal induces the substantial and well-tuned macromolecular forces that the ribosome must generate following every peptide bond formation to trigger the progress of its 25 kDa tRNA substrates through the intersubunit active sites along with its own precise directional movement by one codon down the mRNA.

Here, using optical tweezers, we measure directly the forces exerted between the ribosome and mRNA in the context of various tRNAs before and after peptide bond formation. Ribosome complexes were assembled on a 57 nucleotide, T4 gene 32-derived mRNA containing a natural SD sequence and a 5' biotin modification to tether the complex to streptavidin-derivatized quartz surfaces. For optical trapping, 1μm diameter carboxylate-modified beads coated with anti-digoxigenin antibody were bound to a digoxigenin-modified DNA oligonucleotide designed to hybridize to an extension genetically engineered into helix 44 of the 16S rRNA[16] (Figure. 4a, b). A single bead bound to a single ribosome complex, as verified by single step fluorescence photobleaching of Cy3-labeled tRNA in the P site, was tethered to the surface via mRNA and trapped with optical tweezers (Figure. 4c). The piezo stage was moved at a constant velocity (~100 nm/s) in one direction followed by a movement of the bead (Figure. 4d) until the external force exerted on the ribosome resulted in a rupture event and a rapid return of the bead to the trap center position (Figure. 4e).

3.2. *Rupture force distribution*

3.2.1. *Ribosome complex with mRNA including the SD interaction*

Several examples of rupture force events recorded for tethered mRNA-ribosome-bead complexes are shown in Figure. 5a. Rupture events likely result from ribosome dissociation from the mRNA since no rupture events were observed in control experiments where the ribosomes were crosslinked directly to a cysteine-reactive surface instead of being tethered via mRNA (Figure. 5b) or where the ribosome-mRNA complex was replaced by a surface-attached biotinylated RNA oligonucleotide of the same sequence as the rRNA extension complementary to the DNA oligonucleotide-bead conjugate (Figure. 5c). These results demonstrate that the ribosome-oligonucleotide interactions that anchor the bead onto the ribosome are sufficiently strong for force measurements within the required range. In addition, post-rupture colocalization of the bead and the ribosome complex with fluorescent tetramethylrhodamine-labeled[18] tRNA in the P site suggests that the rupture involves disruption of tRNA-mRNA interactions since after the rupture event the mRNA is attached to the surface while the tRNA remains on the bead-associated ribosome. This provided further proof that the observed rupture does not occur at the ribosome-oligonucleotide-bead linkage.

Binding of tRNAs to the ribosome stabilizes ribosome-mRNA interactions. The rupture force distribution for a ribosome-mRNA complex in the absence of tRNAs showed a single peak at 10.6 pN (Figure. 6a). The addition of $tRNA^{fMet}$ to the P site strengthened the ribosome-mRNA interactions roughly by 5 pN, increasing the rupture force to 15.2 pN (Figure. 6b). Subsequent addition of $Phe\text{-}tRNA^{Phe}$ to the A site resulted in further

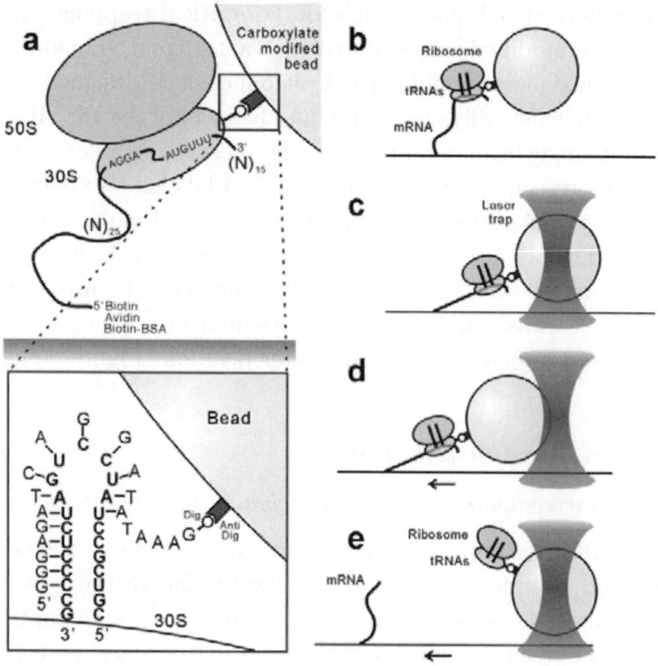

Fig. 4. Experimental design for rupture force measurements on the ribosome. (a) The molecular attachments within the mRNA-ribosome-bead complex. Ribosomal particles were assembled on a short mRNA tethered to the surface via biotin-streptavidin linkage. A digoxigenin-modified oligonucleotide was designed to hybridize to an rRNA loop extension on the small ribosomal subunit. A bead coated with anti-digoxigenin antibody was conjugated to the oligonucleotide and used for optical trapping of the ribosomal complex. (b) The tethered ribosome-bead complex fluctuates around the point of surface attachment. (c) Optical tweezers are used to trap the bead. (d) As the stage with the attached ribosome-bead complex is moved in one direction, the force exerted on the complex increases and the bead becomes displaced. (e) Eventually the external force becomes sufficient to rupture the complex, and the bead returns to the trap center position. (figure taken from S. Uemura et al 2007[17])

stabilization of the complex by another 10 pN, resulting in a single peak distribution centered at 26.5 pN (Figure. 6c). In all experiments, the tRNA occupancy of each ribosome complex was verified just prior to the force measurement using fluorescence resonance energy transfer (FRET). For this purpose, $tRNA^{fMet}$in the P site was labeled with Cy3, Phe-$tRNA^{Phe}$ in the A site was labeled with Cy5, and fluorescence of both dyes was monitored at 532 nm excitation.[19] Only complexes showing FRET due to the presence of both tRNAs were included in further force measurement analysis, so

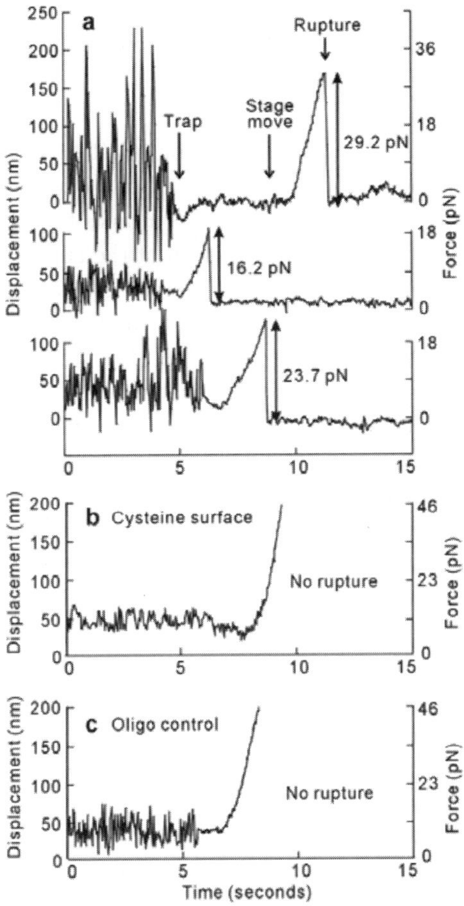

Fig. 5. Examples showing the behavior of the bead. (a) The position of tethered beads fluctuates around the tether center. Once the bead is trapped by the optical tweezers, its fluctuations become suppressed as indicated. As the bead starts to follow the stage movement, the force exerted on the complex increases. When the ribosome complex gets ruptured, the bead returns to the trapping center. (b) Control measurement for ribosomes covalently crosslinked to cysteine-reactive surface. No rupture events were observed within our measurement range. (c) Control measurement for a biotinylated RNA oligonucleotide designed to mimic the extension in the 16S rRNA. The RNA oligonucleotide was attached to streptavidin-derivatized surface and hybridized with the DNA oligonucleotide-bead conjugate. No rupture events were observed. (figure taken from S. Uemura et al 2007[17])

that the final force population distributions are representative of ribosomal complexes with both A and P sites occupied (Figure. 6c-f, i-j). The multiple interactions between P-site and A-site tRNAs and both the large and small ribosomal subunits likely contribute to the stabilization of the ribosomal complex on the mRNA and explain the large rupture force increase upon the addition of tRNAs.

The nature of the aminoacyl group on A-site tRNA strongly affects ribosome-mRNA rupture forces. Binding of the peptidyl-tRNA analogue N-acetyl-Phe-tRNA^{Phe} to the A site, which mimics the post-peptidyl transfer state, resulted in a significant reduction of the rupture force to 12.7 pN (Figure. 6d), suggesting that the interactions between the ribosome and mRNA are weakened after peptide bond formation. Complexes prepared with fMet-tRNA^{fMet} in the P site and Phe-tRNA^{Phe} in the A site, allowing for efficient formation of a ribosome-catalyzed peptide bond, showed a weak rupture force centered at 11.4 pN, with only a minor peak at 24.8 pN most likely representative of a small subpopulation (14.3 %) of either inactive ribosomes and/or ribosomes where the P-site tRNA has become deacylated (Figure. 6e). This provides further evidence that the presence of the peptidyl-tRNA moiety in the A site results in the destabilization of ribosome-mRNA interactions. In addition, the stability of the ribosome on mRNA is not affected by the equilibrium between the classical and hybrid tRNA states. In related work, we have observed that the post-peptide bond mimic (tRNA^{fMet} in the P site and N-acetyl-Phe-tRNA^{Phe} in the A site) results in rapid classical-hybrid state fluctuations at 5 mM Mg^{2+}, (with the occupancy of classical and hybrid states of 3:2) relative occupancy of while the tRNAs remain predominantly in the classical configuration at 15 mM Mg^{2+}.[20] No significant difference was observed in the rupture force distributions of the post-peptidyl transfer complex in 5 mM Mg^{2+} and 15 mM Mg^{2+} buffers (compare Figure. 6d and f). Whereas different Mg^{2+} concentrations can affect tRNA-ribosome complex stability,[21] the timescale of tRNA dissociation is much greater than the time it took to complete the force measurements. The observed lack of strong Mg^{2+} dependence to the measured forces indicates that we are monitoring rupture of mRNA-ribosome base pairings, which are not strongly Mg^{2+} dependent.

3.2.2. *Ribosome complex with mRNA excluding the SD interaction*

The SD interaction directly increases the binding affinity of the ribosome for mRNA and influences tRNA-mRNA translocation.[22] To test the contribution of the SD interaction to complex stability, rupture measurements were performed on ribosomal complexes assembled on an mRNA where the SD interaction had been significantly weakened by modifying the sequence

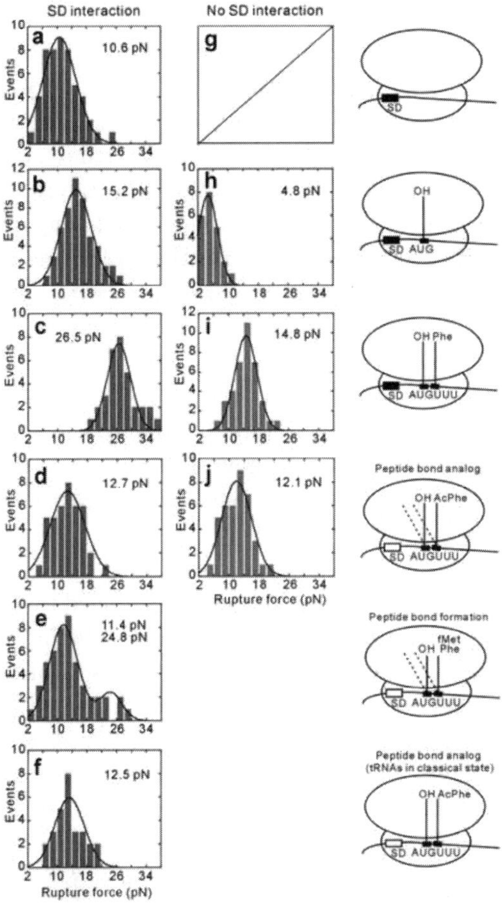

Fig. 6. Rupture force distributions for ribosome complexes assembled on mRNAs containing (a-f) or lacking (g-j) the SD sequence. All complexes were assembled in 5 mM Mg^{2+} unless indicated otherwise. (a, g) Ribosome-mRNA complex without tRNAs. No tethered beads were observed in the absence of the SD sequence (g), indicating that the complex was too unstable for a force measurement under these conditions. (b, h) Ribosome-mRNA complex carrying deacylated $tRNA^{fMet}$ in the P site. (c, i) Ribosome-mRNA complex with $tRNA^{fMet}$ in the P site and Phe-$tRNA^{Phe}$ in the A site. (d, j) Ribosome-mRNA complex with $tRNA^{fMet}$ in the P site and N-acetyl-Phe-$tRNA^{Phe}$ in the A site. (e) Ribosome-mRNA complex after ribosome-catalyzed peptide bond formation. The complex was assembled with fMet-$tRNA^{fMet}$ in the P site and Phe-$tRNA^{Phe}$ in the A site, and incubated for 20 min to allow for peptidyl transfer. (f) Ribosome-mRNA complex with $tRNA^{fMet}$ in the P site and N-acetyl-Phe-$tRNA^{Phe}$ in the A site in 15 mM Mg^{2+}. (figure taken from S. Uemura et al 2007[17])

from AGGA to ACCA.[22] In the absence of tRNAs, no tethered beads were observed indicating that the complexes without the SD region were too unstable for a force measurement (Figure. 6g). The addition of $tRNA^{fMet}$ to the P site stabilized the complexes by almost 5 pN, similarly as for the SD-containing complexes, allowing for a rupture force measurement of 4.8 pN (Figure. 6h). Further addition of Phe-$tRNA^{Phe}$ to the A site increased the rupture force to 14.8 pN (Figure. 6i), a 10 pN increase again similar to that observed for the SD-containing complexes. Upon binding of N-acetyl-Phe-$tRNA^{Phe}$ to the complexes without the SD sequence, a rupture force of 12.1 pN was measured (Figure. 6j), slightly reduced in comparison to the force observed for Phe-$tRNA^{Phe}$ complexes (Figure. 6i). This final post-peptidyl transfer rupture force was essentially the same for the SD-containing and SD-lacking complexes (compare Figure. 6d and j), which further supports the conclusion that peptide bond formation results in destabilization of SD interactions between the mRNA and the small ribosomal subunit.

Although the exact rupture pathway cannot be determined for the different ribosomal complexes, analysis of both the initial and final tRNA occupancy on the ribosomes using fluorescence verified that ribosomal complexes exist in similar ligand-bound states at the beginning and at the end of force measurement experiments. The rupture occurs through the disruption of interactions between the 30S-tRNA complexes and mRNA, upon which the 30S subunit and bound tRNAs remain attached to the bead, while the mRNA stays anchored on the surface via the strong biotin-streptavidin link. Since the initial and final states are comparable for the different complexes, the rupture pathways are also probably very similar and are unlikely to involve an intermediate spontaneous translocation steps[23,24] or dissociation of the entire ribosomal complex on the timescale of the force measurement experiment.

3.3. *Conclusion and discussion*

Our results provide direct evidence for coupling of the 50S peptidyl transferase center and the 30S subunit (Figure. 7). Precisely how the formation of a peptide bond on the 50S portion of the ribosome leads to the weakening of the mRNA contacts with the 30S subunit needed for processive translation is yet to be determined. Previous studies have demonstrated that the ribosome can sense the chemical nature of its ligand in the A site: peptidyl-tRNA shows much lower binding affinity than aminoacyl-tRNA,[21] while deacylation of the A-site bound tRNA can affect the accuracy of translocation along mRNA.[22] Thus, the weaker link between the A-site

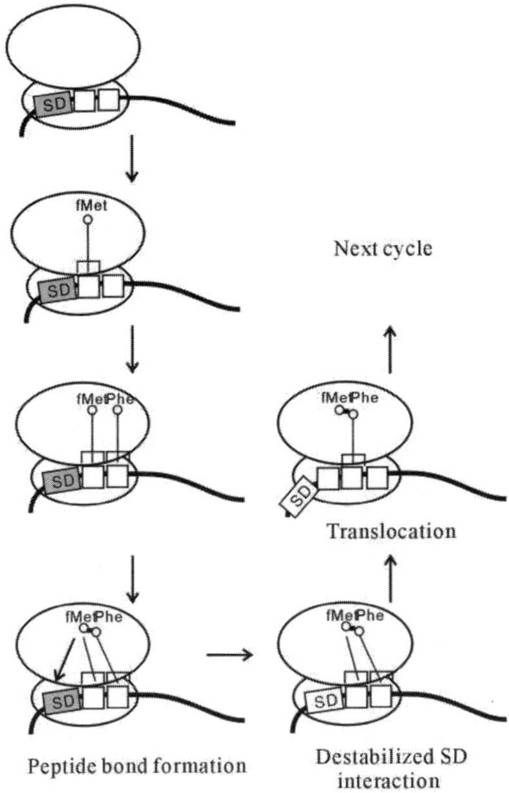

Fig. 7. Ribosome-mRNA initial interaction model. The formation of peptide bond makes SD interaction destabilized to require to facilitate movement of the ribosome along mRNA during the first translocation step.

tRNA and the ribosome may contribute to the decrease in rupture force. The distinct interactions of peptidyl-tRNA versus aminoacyl-tRNA moieties at the peptidyl transferase center after peptide bond formation might induce conformational changes within the 50S subunit, which could be further propagated through the subunit interface towards the anti-SD region of the 30S subunit, possibly via a relative movement of the 50S-30S domains,[22] resulting in the destabilization of SD interactions between the 30S and mRNA, and facilitating the first translocation step. A mechanistic explanation of this allosteric interaction requires further investigation.

Acknowledgments

This work has been done in Steven. Chu (Lawrence Berkeley National Laboratory) group with Joseph. D. Puglisi (Stanford University). S.U. was a recipient of JSPS Postdoctoral Fellowships for Research Abroad. This work was funded by grants to Joseph. Puglisi. from the NIH and the Packard Foundation, to Steven. Chu. from the NSF and NASA, and to Joseph. Puglisi. and Steven. Chu. from the Packard Foundation.

References

1. M. M. Yusupov, et al, *Science* **292**, 883 (2001).
2. G. Z. Yusupova, et al, *Cell* **106**, 233 (2001).
3. D. Moazed and H. F. Noller, *Cell* **57**, 585 (1989).
4. D. Moazed and H. F. Noller, *J. Mol. Biol.* **211**, 135 (1990).
5. S. Yoshizawa, et al, *Science* **285**, 1722 (1999).
6. T. Pape, et al, *EMBO. J.* **18**, 3800 (1999).
7. H. F. Noller, et al, *Science* **256**, 1416 (1992).
8. R. R. Traut and R. E. Monro, *J. Mol. Biol.* **10**, 63 (1964).
9. M. V. Rodnina, et al, *Nature* **385**, 37 (1997).
10. D. N. Wilson, et al, *Curr. Protein Pept. Sci.* **3**, 1 (2002).
11. M. Valle, et al, *EMBO. J.* **21**, 3557 (2002).
12. K. C. Neuman and S. M. Block, *Rev. Sci. Instrum.* **75**, 2787 (2004).
13. A. Ashkin, *Biophys. J.* **61**, 569 (1992).
14. J. Shine and L. Dalgarno, *Proc. Natl. Acad. Sci. USA.* **71**, 1342 (1974).
15. R. A. Calogero, et al, *Proc. Natl. Acad. Sci. USA.* **85**, 6427 (1988).
16. M. Dorywalska, et al, *Nucleic Acids Res.* **33**, 182 (2005).
17. S. Uemura, et al, *Nature* **446**, 454 (2007).
18. M. A. van Dijk, et al, *J. Phys. Chem. B* **108**, 6479 (2004).
19. S. C. Blanchard, et al, *Proc. Natl. Acad. Sci. USA.* **101**, 12893 (2004).
20. H. D. Kim, et al, *Biophys. J.* **93**, 3575 (2007).
21. Y. P. Semenkov, et al, *Nat. Struct. Biol.* **7**, 1027 (2000).
22. K. Fredrick and H. F. Noller, *Mol. Cell* **9**, 1125 (2002).
23. V. I. Katunin, et al, *Biochemistry (Mosc.)* **41**, 12806 (2002).
24. J. Frank and R. K. Agrawal, *Nature* **406**, 318 (2000).

A ROD PROBE REVEALS GAIT OF MYOSIN V

KATSUYUKI SHIROGUCHI

Department of Physics, Faculty of Science and Engineering, Waseda University, Okubo 3-4-1, Shinjuku-ku, Tokyo 169-8555, Japan

Myosin V is a linear molecular motor that moves cargos along actin filaments in a cell. It has two 'feet' (conventionally called 'heads' or 'motor domains'), each attached to a long and relatively stiff 'leg' (traditionally called 'neck' or 'lever arm'), and walks by alternately swinging forward its feet, apparently similar to a human. I describe the mechanisms of how the trailing foot, once lifted, accesses the next forward landing site on actin, which have been revealed by directly observing the leg motion of a walking myosin V through a micrometer-sized biological rod attached to its leg. Next, I discuss our previously proposed mechanism for the lifted foot attaching to the next forward binding site. Finally, I comment on the usefulness of a rod probe as an experimental tool.

1. Linear Molecular Motors and Myosin V

Linear molecular motors, such as myosin, kinesin, or dynein, perform various important roles in a cell, for example, unidirectionally carry cargos along a filamentous track specific for each type of a motor, or maintain the cell shape by producing force while bound to the track. They have a globular motor domain (called "foot" here) that binds to the track and contains the catalytic site for hydrolyzing ATP (adenosine triphosphate), which drives the unidirectional movement.

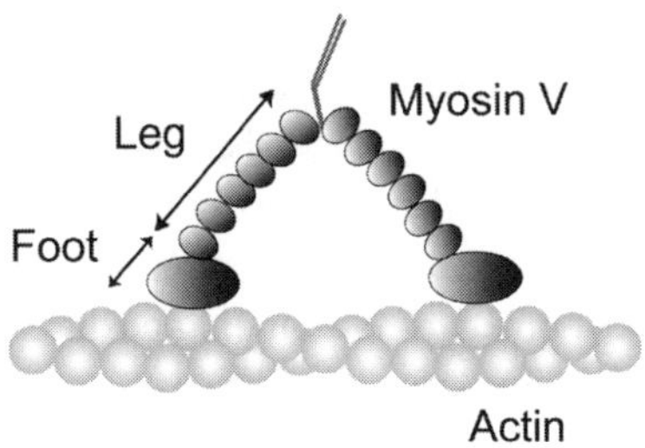

Fig. 1. Myosin V bound to an actin filament (small circles) through the two feet (large ellipsoids). Only the portion of myosin V related to the motile activity is shown. The two legs, each wrapped by six light chains (small ellipsoids), are connected by coiled coil. The distance between the two bound feet is ~36 nm.

Myosins constitute large superfamily, and the structure and molecular properties of each myosin are adapted to perform specific functions in a cell [1, 2]. Myosin V [3, 4] is a transporter, which carries, for instance, vesicles or mRNA along actin filaments. It is a homodimer, having two relatively stiff and long legs connected by a coiled coil. Each leg is wrapped by six light chains (calmodulins) (Fig. 1). Myosin V is one of the most studied linear molecular motors, whose mechanisms are well understood [5, 6].

2. How Myosin V may Walk

Myosin V moves on an actin filament processively [7, 8], which means that a single motor steps along a filamentous track for many ATP hydrolysis cycles without detaching, using its two legs. Several studies have suggested that myosin V 'walks' in discrete ~35-nm steps by alternately swinging its two feet forward [7, 9, 10, 11]. Electron micrographs showed that long legs of myosin V form a V-shape when both feet are attached to actin [12]. By attaching a fluorophore to a leg, the leaning of a leg alternately forward and backward during every step was observed [13, 14]. These features are apparently similar to the human's walking. However, gravity and inertia that affect the walking of a human are negligible in the nano-scale world, and the proteins are subject to an intensive Brownian motion. The walking mechanisms of molecular motors may therefore be different from that of human.

Myosin V has been shown to move unidirectionally without significant backward stepping, at least in the absence of external load [15]. Based on the widely accepted walking model above, myosin V alternately switches between the two- and one-foot binding states [12]. Therefore, there must exist mechanisms for the preferential lifting of a trailing (not leading) foot, and for binding of the lifted foot to a forward site with high success probability. The dissociation of a foot from actin is supposed to be initiated by the ATP binding to its nucleotide-binding site. Therefore, the ADP dissociation from the landed foot, which allows ATP to bind, regulates the dissociation of a foot from actin. The preferential dissociation of the trailing foot was suggested to be due to a change of its ADP affinity depending on the direction-dependent internal strain, which is presumed to exist in a two-foot binding posture [16, 17]. However, the description of how the trailing foot, once lifted, is carried forward to the next binding site has so far been speculative [18].

3. A Micrometer-sized Rod attached to a Leg reveals Walking Mechanism

3.1. *Observing the leg motion in stepping phase by a rod*

Observation of the leg motion during the stepping phase, between the two-foot binding postures, holds a key for understanding how a lifted foot binds to a forward site. To visualize the leg motion in walking myosin V under an optical microscope, we attached a rod (microtubule, see 5.), a few μm in length, to one of the legs (Fig. 2). We expressed a fusion protein, in which calmodulin is genetically connected to a monomeric kinesin mutant [19] that irreversibly binds to a microtubule. Next, the intrinsic calmodulins on the leg were replaced with the fusion protein by changing calcium concentration in solution [8]. Finally, a fluorescently labeled microtubule was added to the myosin V-kinesin mutant complex.

Using optical tweezers, actin filament attached to beads at both ends was suspended, forming an actin bridge, in order to prevent the microtubule from hitting a glass surface (Fig. 2). Driven by the movement of myosin V, the microtubule moved in one direction along the actin bridge, while swinging periodically to-and-fro between two relatively stationary angles separated by ~100°. During the stepping phase, the swing in one direction was always unidirectional, while the swing in the other direction often involved extensive fluctuation before reaching the next stationary angle (Fig. 3b).

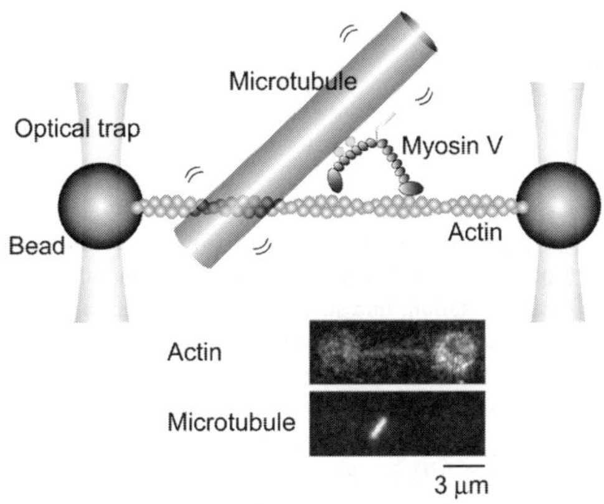

Fig. 2. Actin filament is suspended between two beads trapped by optical tweezers. Microtubule, which serves as a probe, is attached to one of the legs of myosin V through the calmodulin (light chain) - kinesin mutant fusion protein. The legs of myosin and the thickness of the actin filament and microtubule are approximately to scale. Fluorescently labeled actin filament, beads, and microtubule are observed simultaneously in the same view by dual-view microscopy [41].

We also observed the leg motion using as a probe an actin filament (~1 μm in length), in a reversed system where the microtubule was fixed on a glass surface and the actin filament was free (Fig. 3c). The motion of myosin V was essentially the same as in the microtubule-moving assay, and we indeed

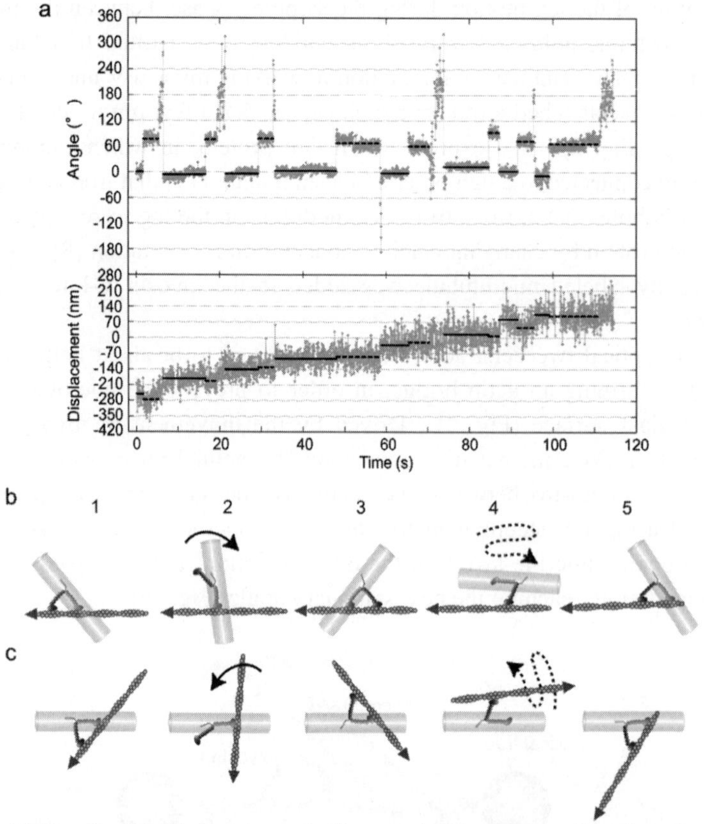

Fig. 3. Motion of a microtubule and an actin filament. (a) Time courses of angular and positional changes. The angle corresponds to a clockwise rotation of a microtubule in the microtubule-moving assay in b (or an anticlockwise rotation, unrestricted in the fluctuation phase, in the actin-moving assay in c). The orientation of the microtubule (or the actin filament) in the position 1 is defined as 0 degrees. Displacements were determined from the position of the center of an actin filament, which basically corresponds to displacement of a microtubule-attached foot along the actin's longitudinal axis in the microtubule-moving assay. Solid and dotted bars, average of the stationary angles. Possible backstepping occurred at ~93 s. Reproduced from Shiroguchi and Kinosita [20] with modifications. (b) Illustration of a microtubule-moving assay. Actin is fixed, and microtubule is moving. Solid arrow, unidirectional swing. Dotted arrow, fluctuation. (c) Illustration of an actin-moving assay. Microtubule is fixed, and actin is moving. Arrowhead at the tip of an actin filament indicates its pointed end. An actin filament translocates in this direction in the assay. Solid arrow, unidirectional swing. Dotted arrow, fluctuation.

observed the movement of actin filament as expected from the microtubule-moving assay, namely, a unidirectional swing of the actin filament and a fluctuating swing between two stationary angles, which occur alternately accompanying translocation (Fig. 3a). (Note that in this actin-moving assay, actin filament translocates in two directions along two stationary angles, as shown in Fig. 3c). Moreover, in order to let actin filament moving freely in three dimensions, we suspended microtubule on large beads immobilized on a glass surface. During fluctuation, we observed moments when a filament became perpendicular to the image plane in which two stationary angles of the filament were primarily located. In the actin-moving assay, we found clearer stepwise translocation of actin, showing that unidirectional swing is not accompanied by any significant translocation of the foot attached to microtubule (1→3 in Fig. 3b), whereas after fluctuating swing the foot takes a ~70 nm forward step from a previous binding site on the actin filament (3→5). These results suggest that the leading leg tilts forward unidirectionally and the trailing leg, once lifted, undergoes extensive fluctuation before landing onto the next binding site ahead.

The measured step size, dwell times of two stationary angles expected to correspond to the ATP-binding rate (see 5.), and the observed angle between two stationary orientations, which correspond to the angle between two legs [20], are consistent with those reported previously [11, 21, 12], which strongly supports the conclusion that we indeed visualized the motion of myosin V's leg.

3.2. *Two mechanisms for the lifted foot to access a forward site*

In the actin-moving assay above, unidirectional swing occurred within 1-4 frames of video record (33 ms per frame) and exhibited no significant backward swing at our time resolution, suggesting that this swing is driven by the active force obtained in ATP hydrolysis cycle. Lever arm (leg) swinging model was originally proposed for myosin II [22], which is a non-processive motor. Myosin II has been shown to drive translocation of actin against a backward force by hydrolyzing ATP [23], and similar results have been reported for a single-leg construct of myosin V [16, 24]. The observed unidirectional swing suggests that the lever arm indeed swings, changing its angle relative to actin, by force (torque) production.

The fluctuating swing took ~0.6 s on average, but sometimes was shorter (e.g., at ~87 s in Fig. 3a) or longer (e.g., at ~72 s in Fig. 3a) in the actin-moving assay. These apparent stochastic motions are suggested to be basically rotational Brownian motion by mean-square-angle-change analysis [20]. Dunn

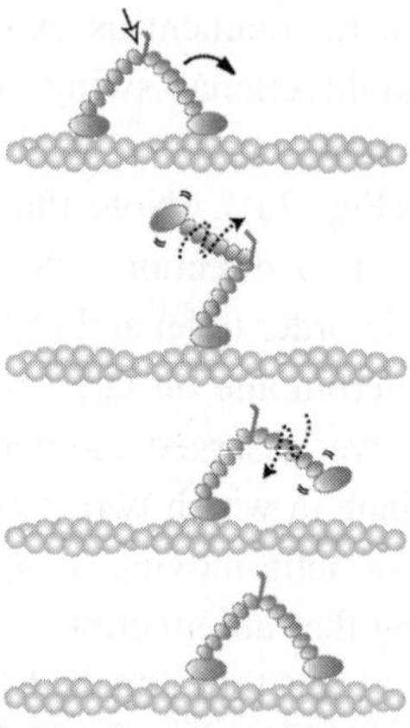

Fig. 4. 'Walking' model of myosin V. Once the trailing foot is lifted, the landed leg tilts (solid arrow) and moves a free joint (open arrow) forward, whereas the lifted leg undergoes rotational Brownian motion (dotted solid arrows) around the joint. Since the joint is biased forward, the lifted foot can access the next forward binding site on actin.

and Spudich [25] attached a 40-nm gold particle to a leg and showed that the particle undergoes translocational diffusion, presumably during the stepping phase. Our results are consistent with their observation and further indicate that diffusion is primarily due to rotational motion of the leg in all directions. The free diffusion of a lifted foot is consistent with the presence of a free joint at the leg-leg junction, reported earlier for myosin II [26] and myosin V [12].

The observed two types of a swing thus suggest the following: the lifted leg, which is a trailing leg having dissociated from actin, undergoes Brownian rotation around a flexible joint at the leg-leg junction as a pivot, whereas the active lever action of the landed leg drives the pivot forward, and eventually the lifted foot lands onto the next forward binding site (Fig. 4). Therefore, myosin V walks by the combination of two processes, the ATP-powered lever action and the Brownian motion.

4. On the Toe Up-down Mechanism for the Lifted Foot binding to a Forward Site

We have previously proposed another mechanism for the two-foot linear molecular motors, which we call the toe-down mechanism, for the binding of the lifted foot to a forward site [27, 28, 29]. Electron microscopic studies of myosin V have shown that there is no significant difference in foot orientation on an actin filament between the leading and the trailing feet [12, 30], indicating that the 'sole' (actin binding site) of the foot is parallel to an actin filament while bound to it. Therefore, during the fluctuating forward swing, the ankle of the lifted foot may be reoriented to take a toe-down posture, at least just before the landing. If the toe in the lifted foot is being pointing down, it ensures that the lifted foot binds to a forward site. In this case, myosin V would move forward even against backward force, which weakens the effect of the forward bias by the lever action of the landed foot.

An interesting question is how this toe up-down (the ankle angle) movement and the related affinity for actin are regulated during the fluctuating swing. The ankle motion may be coupled to a nucleotide binding to a foot. It is thought that the ATP binding to the trailing foot induces its detachment from actin, and the phosphate (Pi) release from the ADP-Pi state occurs just after the lifted foot (re)binds to actin. Therefore, we focus here on the difference between the orientation of a foot relative to a leg in the ATP and ADP-Pi states in the absence of actin. An electron microscopic study of the negatively stained myosin V [30] indicates that in the nucleotide-free state and the steady-state ATP hydrolysis state (primarily ADP-Pi state) in the absence of actin the ankles exhibit the orientations differing by ~90°. Since the nucleotide-free state is thought to correspond to the toe-up posture of the trailing foot, the toe may be down in the ADP-Pi state. A crystal structure of myosin V with MgADP.BeFx bound at the active site, suggested to correspond to the ATP-bound state, appears to show the toe-up posture [31]. These studies suggest that the toe-up and the toe-down postures correspond to the ATP and the ADP-Pi states, respectively. In addition, two FRET (fluorescence resonance energy transfer) solution studies using myosin II have suggested that the ADP-Pi state corresponds to the toe-down posture [32, 33].

It is thought that there exists equilibrium between the ATP and ADP-Pi states of myosin, which means that the ankle orientation might vary between the toe-up and toe-down postures during the fluctuating swing. However, even if the toe-up and the toe-down postures occur with the same probability, the higher affinity for binding to a forward site in the toe-down posture compared to the affinity for binding to a backward site in the toe-up posture increases the possibility that the lifted foot binds to a forward site. According to the above studies so far, the lifted foot in the toe-down posture has ADP-Pi bound, which means that the foot in the toe-down posture has high affinity for a forward site. In addition, in this process, the stiffness of the ankle and the leg is important. If the ankle and/or leg of the ADP-Pi bound lifted foot are flexible to some extent, the foot, having high affinity for actin, may land at the same site on actin from which it had just dissociated, especially in the presence of backward load. In this case, one ATP may be consumed without stepping forward. Thus, the characterization of the toe up-down motion in the lifted foot may be another key to fully reveal the mechanism of the unidirectional movement of myosin V.

5. Biological Rod as a Probe and its Attachment

The microtubule bound to a leg does not seem to affect at least the ATP-waiting time of the foot. In the actin-moving assay (see 3.1), the dwells before unidirectional swing (7.7 ± 0.5 s) and fluctuating swing (7.1 ± 0.9 s) at 0.2 μM ATP are similar to 5.6 s that is calculated from the ATP-binding rate of 0.9 $\mu M^{-1}s^{-1}$ for myosin V [21]. A micron-sized rod slows down the motion of proteins due to increased viscous load, but the essential features are most likely preserved, as shown for the rotary motor F_1-ATPase [34]. Looking at this from a different perspective, the slowed motion of the rod against the viscous load allows to infer the force behind the motion, such as active force or Brownian motion, as done in 3.2. The rod is an especially useful tool for the detection of the angle change, which also means that it may be used not only for observing the linear movement, for example, that of the molecular motors, but also for monitoring the structural changes in proteins accompanied by the angle shift in some parts. One of the easiest ways to understand a feature of a structural change is to look at the rod motion through a microscope with your eyes, which gives the most incontrovertible comprehension and which helps to easily understand the results for non-experts and even non-scientists. To attach a rod to the object of your interest might occasionally be challenging, but if it has worked, you will be fascinated by its motion.

An actin filament has been used to show rotation of F_1-ATPase under an optical microscope [35]. Here we used both a microtubule and an actin filament to observe the leg motion of myosin V [20]. Very recently, by observing an actin filament, Komori *et al.* [36] monitored the foot orientation of myosin V. Actin and microtubule are useful because their properties have been well studied. It is easy to label and modify them to attach to a specific point of interest, and their size can be changed by simply intensively mixing the solution, for example, using a pipette.

We attached microtubule using a fusion protein (as described in 3.1.). Yajima *et al.* [37] fused kinesin with gelsolin that severs an actin filament and caps its barbed end in the presence of Ca^{2+}. A single kinesin was labeled with multiple fluorophores by adding fluorescently labeled actin filament, and its movement along a microtubule was observed for a long time even at low ATP concentrations. In addition, this system has been modified [38] by using Ca^{2+}-

insensitive gelsolin mutant [39, 40], as microtubules are unstable in the presence of Ca^{2+}. By genetically connecting a calmodulin-binding sequence (e.g., a part of myosin V leg) to a target protein, one can attach a rod (microtubule) to the protein and the attachment can be regulated by calcium concentrations as done in 3. Originally, the use of fusion proteins is well known to be one of most useful biochemical techniques, as, for example, GST-fusion protein has been used for protein purification for years. A fusion protein is applicable for attachment of an observable protein to your target protein or its potion, which allows to directly observe the motion of (a part of) protein under a microscope.

Acknowledgments

I thank Kazuhiko Kinosita, Jr. for encouragement and insightful advices, and Kinosita lab members for discussions for the work in 3. I am also grateful to Sergey V. Mikhailenko and Tetsuaki Okamoto for critical reading of the manuscript and discussions, and to Yusuke Oguchi, Makito Miyazaki, and Yuji Tajima for comments. This work was supported in part by Grants-in-Aid for Special Purposes from the Ministry of Education, Culture, Sports, Science and Technology of Japan.

References

1. V. Mermall, P. L. Post, M. S. Mooseker, *Science* **279**, 527-33 (1998).
2. R. D. Vale, *Cell* **112**, 467-80 (2003).
3. E. M. Espreafico et al., *J. Cell Biol.* **119**, 1541 (1992).
4. R. E. Cheney et al., *Cell* **75**, 13 (1993).
5. R. D. Vale, *J. Cell Biol.* **163**, 445 (2003).
6. J. R. Seller and C. Veigel, *Curr. Opin. Cell Biol.* **18**, 68 (2006).
7. A. D. Mehta et al., *Nature* **400**, 590 (1999).
8. T. Sakamoto, I. Amitani, E. Yokota and T. Ando, *Biochem. Biophys. Res. Commun.* **272**, 586 (2000).
9. R. Matthias et al., *Proc. Natl. Acad. Sci. USA* **97**, 9482 (2000).
10. M. Y. Ali et al., *Nat. Struct. Biol.* **9**, 464 (2002).
11. A. Yildiz et al., *Science* **300**, 2061 (2003).
12. M. L. Walker et al., *Nature* **405**, 804 (2000).
13. J. N. Forkey, M. E. Quinlan, M. A. Shaw, J. E. Corrie and Y. E. Goldman, *Nature* **422**, 399 (2003).
14. E. Toprak et al., *Proc. Natl. Acad. Sci. USA* **103**, 6495 (2006).
15. D. M. Warshaw et al., *Biophys. J.* **88**, L30 (2005).

16. C. Veigel, S. Schmitz, F. Wang and J. R. Sellers, *Nat. Cell Biol.* **7**, 861 (2005).
17. T. Purcell, H. L. Sweeney and J. A. Spudich, *Proc. Natl. Acad. Sci. USA* **102**, 13873 (2005).
18. C. Veigel, F. Wang, M. L. Bartoo, J. R. Sellers and J. E. Molloy, *Nat. Cell Biol.* **4**, 59 (2002).
19. I. M. Crevel et al., *EMBO J.* **23**, 23 (2004).
20. K. Shiroguchi and K. Kinosita, Jr., *Science* **316**, 1208 (2007).
21. E. M. De La Cruz, A. L. Wells, S. S. Rosenfeld, E. M. Ostap and H. L. Sweeney, *Proc. Natl. Acad. Sci.USA* **96**, 13726 (1999).
22. H. E. Huxley, *Science* **164**, 1356 (1969).
23. J. T. Finer, R. M. Simmons and J. A. Spudich, *Nature* 368, 113 (1994).
24. J. R. Moore, E. B. Krementsova, K. M. Trybus and D. M. Warshaw, *J. Cell Biol.* **155**, 625 (2001).
25. A. R. Dunn and J. A. Spudich, *Nat. Struct. Mol. Biol.* **14**, 246 (2007).
26. K. Kinosita, Jr., S. Ishiwata, H. Yoshimura, H. Asai and A. Ikegami, *Biochemistry* **23**, 5963 (1984).
27. M. Y. Ali et al., *Biophys. J.* **86**, 3804 (2004).
28. K. Kinosita, Jr., M. Y. Ali, K. Adachi, K. Shiroguchi and H. Itoh, *Adv. Exp. Med. Biol.* **565**, 205 (2005).
29. K. Kinosita, Jr., K. Shiroguchi, M. Y. Ali, K. Adachi and H. Itoh, *Adv. Exp. Med. Biol.* **592**, 369 (2007).
30. S. Burgess et al., *J. Cell Biol.* **159**, 983 (2002).
31. P.-D. Coureux, H. L. Sweeney and A. Houdusse, *EMBO J.* **23**, 4527 (2004).
32. Y. Suzuki, T. Yasunaga, R. Ohkura, T. Wakabayashi and K. Sutoh, *Nature* **396**, 380 (1998).
33. W. M. Shih, Z. Gryczynski, J. R. Lakowicz and J. A. Spudich, *Cell* **102**, 683 (2000).
34. R. Yasuda, H. Noji, M. Yoshida, K. Kinosita, Jr. and H. Itoh, *Nature* **410**, 898 (2001).
35. H. Noji, R. Yasuda, M. Yoshida and K. Kinosita, Jr., *Nature* **386**, 299 (1997).
36. Y. Komori, A. H. Iwane and T. Yanagida, *Nat. Struct. Mol. Biol.* **14**, 968 (2007).
37. J. Yajima, M. C. Alonso, R. A. Cross and Y. Y. Toyoshima, *Curr. Biol.* **19**, 301 (2002).
38. K. Shiroguchi, M. Ohsugi, M. Edamatsu, T. Yamamoto and Y. Y. Toyoshima, *J. Biol. Chem.* **278**, 22460 (2003).
39. A. Lueck, H. L. Yin, D. J. Kwiatkowski and P. G. Allen, *Biochemistry* **39**, 5274 (2000).
40. K. M. Lin, M. Mejillano and H. L. Yin, *J. Biol. Chem.* **275**, 27746 (2000).
41. K. Kinosita, Jr. *et al., J. Cell Biol.* **115**, 67 (1991).

MECHANISM OF SPONTANEOUS OSCILLATION EMERGING FROM COLLECTIVE MOLECULAR MOTORS

YUTA SHIMAMOTO

SHIN'ICHI ISHIWATA

Department of Physics, Faculty of Science and Engineering, Waseda University, 3-4-1 Okubo, Shinjuku-ku, Tokyo 169-8555, Japan

Biological systems include a large number and various kinds of molecular machines. Individual molecular machines work stochastically, while the systems constructed of the ensembles of these machines exhibit dynamically-ordered phenomena, rather than a simple sum of individual parts. Here we focus on the spontaneous oscillatory contraction (SPOC) observed in the contractile system of muscle. From the mechanical measurements in the precursor state of SPOC, we discuss how the functions of individual molecular motors are autonomously regulated in the contractile system.

1. Introduction

The contractile system of muscle mainly consists of the linear force-generating motors, myosins, and their counterpart, actin. Both myosin and actin form the thick and thin filaments, respectively, which are further organized in the bipolar three-dimensional liquid crystalline-like lattice structure, termed sarcomere (Fig. 1A). The sarcomeres are connected in series to form a myofibril, and the myofibrils are connected in parallel to each other to form the muscle fiber. The contractile force originates from the collective power strokes of individual myosin heads stochastically interacting with actin, which results in sliding of these two filaments relative to each other, utilizing the energy of ATP hydrolysis. The frequency of force generation by each myosin head depends on the level of thin filament activation, which is controlled by $[Ca^{2+}]$ in the surroundings.

Under typical physiological conditions muscles generate constant force and movement. On the other hand, it is known that skinned striated muscle exhibits steady periodic oscillation of both force and sarcomere length, named SPOC (Spontaneous Oscillatory Contraction), when the state of the contractile system is intermediate between contraction and relaxation.[1] During SPOC each sarcomere periodically repeats a slow shortening phase and a rapid lengthening phase with the peak-to-peak amplitude of >200 nm, which is significantly larger than the stroke size of a myosin head. In addition, the lengthening phase is transmitted to the adjacent sarcomeres along the long axis of the myofibril (Fig. 1B). These phenomena strongly suggest that the contractile system itself possesses a self-regulatory mechanism in addition to being passively regulated by Ca^{2+}.

It is expected that the understanding of the molecular mechanism of SPOC leads to reveal how the stochastic processes of individual reactions are self-regulated in biological systems (Fig. 1C). Under the SPOC condition the thin filament is activated by myosin heads strongly bound to the thin filament in the nucleotide-free and ADP-bound states independently of Ca^{2+},[2] suggesting that myosin heads play a crucial role in the self-regulation of contractility. Nevertheless, the mechanism of self-regulation underlying SPOC is difficult to understand because of its complex and cooperative nature, which includes multiple mechano-chemical feedback loops.

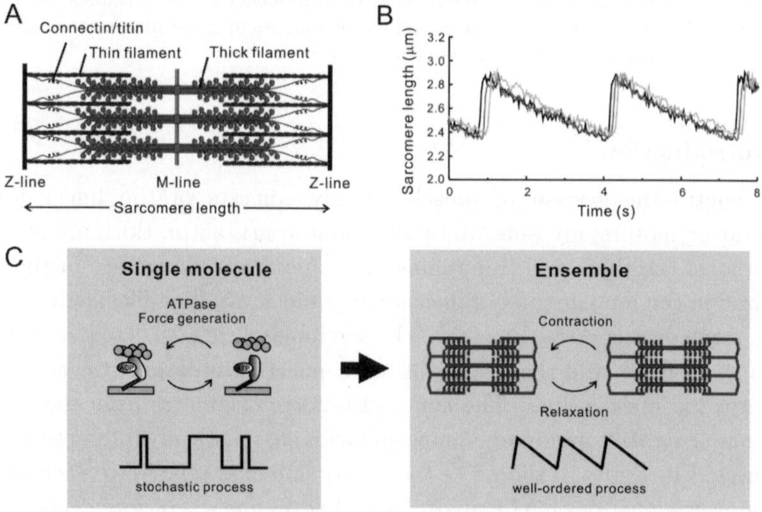

Fig. 1. (A) The structure of a sarcomere. (B) The waveform of three adjacent sarcomeres during SPOC, connected in series along a myofibril. (C) How does an ensemble of stochastic molecular motors generate well-ordered phenomena?

In this paper we discuss the molecular mechanism of SPOC based on the mechanical properties of a sarcomere at the intermediate levels of activation that we clarified recently.[3,4]

2. Mechanical Measurements in Myofibrils

To reveal the mechanism of self-regulation underlying SPOC, we investigated the length-dependent properties of force development in sarcomeres, mostly in the precursor state to SPOC, known as the ADP-induced contraction conditions.[3] SPOC is induced by the addition of inorganic phosphate to these conditions. A single rabbit glycerinated skeletal (psoas) myofibril (or their thin bundle) containing 10-20 sarcomeres connected in series was suspended between a pair of thin glass microneedles, one of which was >50-fold stiffer than the other, on an inverted phase-contrast microscope. The myofibrils in relaxing solution (1 mM MgATP, 2 mM free Mg^{2+}, 4 mM EGTA, 20 mM MOPS (pH 7.0), and 1 mM DTT) were activated by abruptly exchanging the solution to the ADP-induced contraction solution containing high concentrations of [MgADP] (1 mM MgATP, 4-20 mM MgADP, 2 mM free Mg^{2+}, 4 mM EGTA, 20 mM MOPS (pH 7.0), and 1 mM DTT), which allows examining the mechanism of regulation mediated by the ADP-bound myosin heads separately from the regulation by Ca^{2+}. The ionic strength of either solution was adjusted by KCl to 180 mM. The sarcomeres were homogeneously activated/relaxed with a dual laminar flow system.[5] All experiments were carried out at $23 \pm 1\ ^\circ$ C.

The myofibrillar force was estimated from the deflection of the flexible needle, the stiffness of which was 0.1-1.0 μN/μm. The maximum active force was calculated by subtracting the resting force from the total (resting + active) force at the steady state of force generation. The individual sarcomere lengths were calculated from the intensity profile of the phase-contrast image of the myofibril. For details, see ref. 3.

3. Mechanical Properties of Sarcomeres at the Precursor State to SPOC

3.1. *Cooperative force generation modulated through the sarcomere structure*

It has been established that the active isometric force of striated muscle is proportional to the length of overlap between the thick and thin filaments when muscle is fully activated. Hence, the active force linearly decreases with increasing sarcomere length accompanied by a decrease in the length of overlap (broken line in Fig. 2B).[6] This linear relationship between the active force and the number of myosin heads provides a fundamental basis that each myosin head works as an independent force generator in muscle. On the other hand, at the precursor state to SPOC achieved with 1 mM MgATP and 4 mM MgADP in the absence of Ca^{2+}, we found that the larger active force was generated at longer sarcomere length, at the region of sarcomere length between 2.2 and 3.0 μm (Fig. 2A).[3] That is, the activity per a unit length of overlap enhanced with increasing sarcomere length, suggesting that the contractile system itself possesses a length-sensing mechanism.

It has been reported that an increase in sarcomere length causes a reduction of the interfilament spacing between the thick and thin filaments, because connectin/titin, which contributes to the resting force independently of the interaction between the thick and thin filaments, extends from the Z-line to the M-line in sarcomeres[7] (see Fig. 1A). This reduction of the interfilament spacing may cause the larger force production at longer SL due to an increase in the effective concentration of myosin heads in the vicinity of the thin filament. We found that the osmotic compression of the interfilament spacing by $\leqslant$1 nm, induced by the addition of 1% dextran T-500, enhanced the active force two-fold at the sarcomere length of 2.3-2.6 μm (Fig. 2A). The active force increased with an increase in the dextran concentration up to 2%, while further addition of dextran caused the reduction of force.[3] These results strongly suggest that there exists an optimal interfilament spacing for the interaction of myosin heads with the thin filament, and therefore the interfilament spacing is critical for the sarcomere length-dependent activation.

3.2. *Cooperative force generation through the myosin binding*

An increase in [MgADP] from 4 mM to 20 mM prominently altered both the active force and its length-dependence (Fig. 2B). At 4 mM MgADP the active force substantially increased with increasing sarcomere length from 2.2 to ~3.0 μm and monotonically decreased with a further stretch of sarcomeres (open circles in Fig. 2B). As [MgADP] increased, the sarcomere length at which the maximal active force was generated gradually shifted leftward, accompanied by an increase in the magnitude of active force. Concomitantly, the region of sarcomere lengths where the active force monotonically decreased with increasing sarcomere length became broader, approaching the well-established linear force-length relationship obtained at the maximal activation by Ca^{2+} (broken line in Fig. 2B). If the length-dependent change in the interfilament spacing is predominant regardless of the level of activation, the active force must increase with increasing

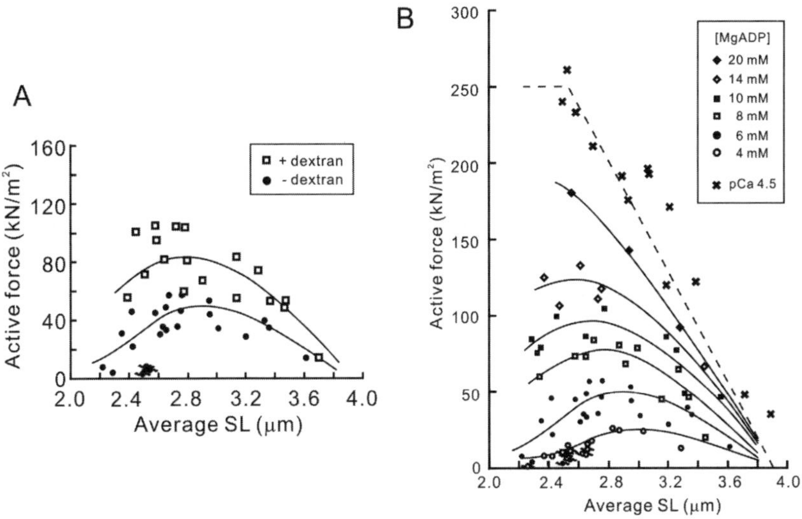

Fig. 2. Force-length relationship at the precursor state to SPOC (A) and at various levels of activation by MgADP (B). Data were taken from Shimamoto et al,[3] with modifications. The plots with wavy line indicate where the SPOC was observed.

sarcomere length even at maximal activation. However, the force enhancement with increasing sarcomere length diminished as the level of activation increased. Therefore, the contribution of the interfilament spacing effect must be reduced with an increase in the density of bound myosin heads. Considering that an increase in [MgADP] has the effect similar to the osmotic compression of the interfilament spacing, we infer that the binding of myosin heads to the thin filament results in the spatial modulation of the relative distance between the thick and thin filaments. That is, each myosin head works as a steric modulator in addition to its role as a force generator so that cooperative activation occurs through the modulation of the lattice structure of a sarcomere. This effect is well explained by a simple sarcomere model, in which the relative distance between the myosin heads and the thin filament varies depending on the density of myosin heads bound to the thin filament.[3]

3.3. *Cooperative force generation under various activating states*

The larger force production at longer sarcomere lengths as shown above can be obtained at different activating conditions. This type of length-dependent cooperative force generation was first found by Endo[8] and recently confirmed in our work[4] at partial activation by Ca^{2+}. We found that it also occurs not only at the ADP-induced activation as described above, but at the rigor-activation (pMgATP = 4.7, 2 mM free Mg^{2+}, 4 mM EGTA, 20 mM MOPS (pH 7.0), and 1 mM DTT) as well, where the sarcomeres are activated by an increase in the population of nucleotide-free myosin heads with lowering [MgATP] in the absence of either Ca^{2+} or MgADP. Fabiato and Fabiato previously reported that at a rigor-activation the force-length relationship is linear.[9] However, when the level of activation is sufficiently low, we found that the relationship becomes nonlinear (Fig. 3), being similar to that observed at the intermediate levels of activation by either Ca^{2+} or MgADP. It is to be noted here that the relative populations of the nucleotide-bound states of the collective myosin motors differ between the activating conditions. Therefore, it may be inferred that the critical factor for the cooperative activation is not the relative proportion of the nucleotide states of myosin heads, but the balance between the strong-binding and weak-binding myosin heads, which governs the length-dependent changes in the interfilament spacing.

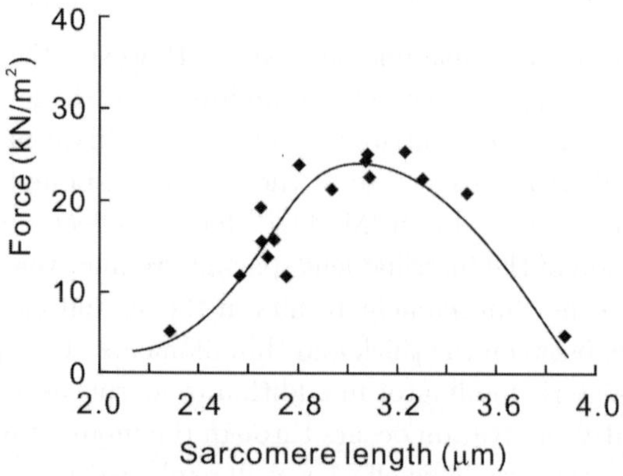

Fig. 3. Force-length relationship at an activation induced by rigor myosin heads at pMgATP = 4.7 in the absence of either Ca^{2+} or ADP. Data were obtained from 3 myofibrils.

4. Molecular Mechanism of SPOC

We demonstrated that at the precursor state to SPOC the activity of a unit length of overlap between the thick and thin filaments strongly depends on sarcomere length, that is, the larger active force is produced when sarcomeres are stretched over its natural length of ~2.4 μm (Fig. 2A). In addition, the high sensitivity of active force generation to the osmotic compression of the lattice structure (Fig. 2A) suggests that the mechanism is, at least in part, attributable to the changes in the interfilament spacing between the thick and thin filaments, induced by the changes in sarcomere length.

When muscle is maximally activated, longer sarcomeres become unstable compared with shorter ones, because the active force decreases with increasing sarcomere length, and therefore, the lengthened sarcomeres continue to lengthen due to larger force exerted by shortening sarcomeres connected in series along the myofibril. At the precursor state to SPOC, however, the larger active force was generated at longer sarcomere length, unlike that at maximal activation. This property must contribute to the stability of oscillation during SPOC, where the sarcomeres tend to become unstable and stable, respectively, as the shortening and lengthening proceeds, because of the expansion and compression of the lattice structure of sarcomeres.

Besides, as we mentioned above, the lattice structure must be modulated when activated, to account for the [MgADP]-dependent changes in the force-length relationship (Fig. 2B). Based on this property, it is expected that there exists bimodal lattice structure at the same sarcomere length depending on whether the myosin heads are bound to the thin filament or not. That is, the structure of a sarcomere may be compressed when the myosin heads attach to the thin filament, compared with the relaxing conditions where myosins only weakly interact with the thin filament. This idea is supported by the X-ray diffraction studies.[10] During SPOC the sarcomeres generate active force at the shortening phase, while they yield to the external load at the lengthening phase. This indicates that the sarcomeres adopt a compressed lattice structure when shortening (the states 1 & 4 in Fig. 4) and an expanded lattice structure when lengthening (the states 2 & 3 in Fig. 4).

Besides, our preliminary observations showed that under intermediate levels of activation the force-generating sarcomeres yield to a stepwise increase in load to a few percents above the maximal active force. It is considered that the stability of a sarcomere is determined by the balance between the attachment of myosin heads to and their detachment from the thin filament. When this balance shifts to the detachment side, it leads to the transient relaxation accompanied by the cooperative detachment of myosin heads from the thin filament, because the bond lifetime of each myosin head exponentially decreases with an increase of load exerted per myosin head.[11]

During SPOC this distortion possibly results from the number fluctuation of force-generating myosin heads,[12] the reversed power stroke of myosin heads induced by phosphate binding,[13] and the tug-of-war between sarcomeres being connected in series and oscillating out of phase. Therefore, the switching between two different states of a sarcomere is regulated by the force balance between the active force and the external load.

Based on the above considerations, we propose that the mechanism of the dynamic oscillation of sarcomeres emerges as follows (Fig. 4). A sarcomere tends to be destabilized due to the expansion of the interfilament spacing as the shortening proceeds (from the state 4 to 1). A disruption of force balance induces cooperative unbinding of myosin heads, resulting in the expansion of the compressed lattice structure (from the state 1 to 2). Once the thin filament separates further from the myosin heads, the probability of the interaction between them prominently decreases, leading to the transient relaxation (from the state 2 to 3). Since the lattice spacing becomes narrower as the lengthening of sarcomeres proceeds, the active force starts to recover, accompanied by the reattachment of myosin heads (from the state 3 to 4), and the shortening starts to occur when the active force exceeds the external load.

In conclusion, we propose that the dynamic cooperative motion of the collective molecular motors in muscle is spontaneously generated by the self-regulation of motor activity through the length-dependent modulation of the three-dimensional lattice structure of sarcomeres and the applied load.

5. Summary

The mechanical measurements at the precursor state to SPOC revealed the self-regulatory mechanisms underlying the performance of the collective molecular motors in muscle. The cooperative motion of the stochastic molecular motors is achieved by the hierarchical feedback, such that the force generation by motors modulates the sarcomere structure, and the modulation of the sarcomere structure in turn regulates each motor's activity. This implies that in biological systems the self-regulation of reactions is performed not only by the direct interaction between functional molecules, but by the organized higher-order structure and the strain as well. The molecular mechanism of SPOC will push forward the understanding of the higher-ordered regulatory mechanisms that the biological systems possess.

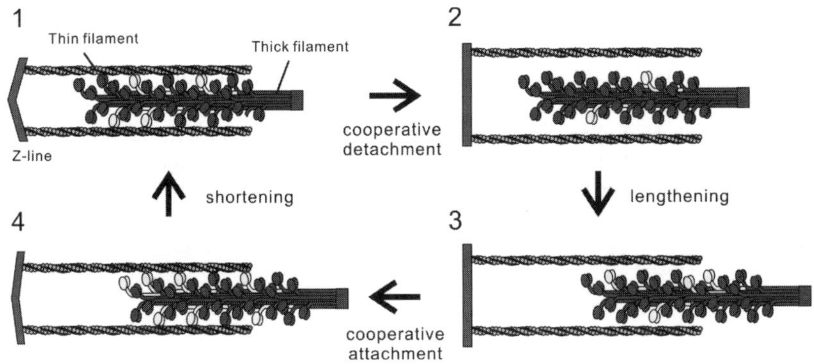

Fig. 4. Illustration of the molecular mechanism of SPOC through the lattice structure of a sarcomere. Only the left half-sarcomere is shown. For simplicity, thin filaments are represented as a rigid rod. The neutral position of the thin filament on Z-line changes depending on sarcomere length. The restoring force is assumed to be stored in the Z-line structure, through the modulation of the lattice structure by cross-bridge formation.

Acknowledgments

We thank Drs. Chiara Tesi (Universita di Firenze), Junichi Wakayama (National Food Research Institute) and Madoka Suzuki (Waseda University) for their technical advice on the mechanical measurements on myofibrils. We thank Dr. Sergey V. Mikhailenko (Waseda University) for his critical reading of the manuscript. This work was supported by Grants-in-Aid for Specially Promoted Research, the 21st Century COE Program, Scientific Research (A), and "Academic Frontier" Project (to S.I.) from the MEXT, Japan, and Waseda University Grant for Special Research Projects (to S.I. and Y.S.).

References

1. N. Okamura and S. Ishiwata, *J Muscle Res Cell Motil.* **9**(2), 111 (1988)
2. H. Shimizu, T. Fujita, and S. Ishiwata, *Biophys J.* **61**(5), 1087 (1992)
3. Y. Shimamoto, F. Kono, M. Suzuki, and S. Ishiwata, *Biophys J.* **93**, 4330 (2007)
4. Y. Shimamoto, M. Suzuki, and S. Ishiwata, *Biochem Biophys Res Commun.* **366**(1), 233 (2008)
5. F. Colomo, S. Nencini, N. Piroddi, C. Poggesi, and C. Tesi, *Adv Exp Med Biol.* **453**, 373 (1998)
6. A.M. Gordon, A.F. Huxley, and F.J. Julian, *J Physiol.* **184**(1), 170 (1966)
7. N. Fukuda and H.L. Granzier, *J Muscle Res Cell Motil.* **26**(6-8), 319 (2005)
8. M. Endo, *Nat New Biol.* **237**(76), 211 (1972)
9. A. Fabiato and F. Fabiato, *J Gen Physiol.* **72**(5), 667 (1978)

10. I. Matsubara, Y.E. Goldman, and R.M. Simmons, *J Mol Biol.* **173**(1), 15 (1984)
11. T. Nishizaka, H. Miyata, H. Yoshikawa, S. Ishiwata, and K. Kinosita Jr, *Nature.* **377**(6546), 251 (1995)
12. M. Suzuki, H. Fujita, and S. Ishiwata, *Biophys J.* **89**(1), 321 (2005)
13. J.A. Dantzig, Y.E. Goldman, N.C. Millar, J. Lacktis, and E. Homsher, *J Physiol.* **451**, 247 (1992)

SIMULATED ROTATIONAL DIFFUSION OF F_O MOLECULAR MOTOR

HISATSUGU YAMASAKI* and MITSUNORI TAKANO†

Department of Physics, Waseda University,
3-4-1 Okubo, Tokyo 169-8555, Japan
**E-mail: hisa@tb.phys.waseda.ac.jp, †E-mail: mtkn@waseda.jp*

F_O is a membrane-embedded part of ATP synthase, transforming proton current running within this molecule into a rotational motion taking place between two subunits, c-subunit and a-subunit. How the proton current could be transformed into the rotational motion remains unclear. In this paper, by means of molecular dynamics simulation, we studied basic properties of the rotational motion. In equilibrium, the c-ring, which consists of 10-12 c-subunits and interacts with the a-subunit, showed stepwise rotational motion. The rotational motions resulted in a free rotational diffusion in a longer period. The diffusion constant calculated as a function of temperature showed a glass-transition-like behavior: at the lower temperatures, diffusional motion was significantly suppressed, deviating from the Einstein's relation. Under a non-equilibrium condition where different heat baths with different temperatures are applied, respectively, to the c-ring and the a-subunit, we found that directionality arises in the rotational diffusion. We finally point out how the structural flexibility (i.e., softness) of protein molecules pertains to our results.

Keywords: MD simulation; molecular motor; Feynman ratchet; proton channel.

1. Introduction

In living organisms, F_OF_1-ATP synthase produces ATP (adenosine triphosphate) by using the proton current down the electrochemical potential maintained across the cell membrane. The F_OF_1-ATP synthase (Fig.1) consists of two main parts: one part, F_O, is embedded in the membrane and contains proton channels, and the other part, F_1, protrudes into the inner region of the cell and catalyzes ATP synthesis and hydrolysis.

In the F_O part, the proton current running through the putative proton channels within this molecule is thought to generate a rotary motion between the c-subunit and the a-subunit. How the proton current generates the rotary motion is not yet clear. Elucidating the proton pathway in F_O, therefore, is one of the most crucial point in understanding the mechanism of the rotary motion generation.[1] Recent theoretical models based on the so-called "half-channel model"[1] have presented plausible explanation of the rotary motion generation.[2,3] They are, however, basically low dimensional reaction-diffusion models, and hence are far from realistic, even though they might capture some essential aspects of the mechanism of the rotary motion.

We therefore constructed a more realistic model of the F_O motor, particularly considering the following three aspects: 3D structure of protein, flexibility of the structure, and the physico-chemical interaction between molecules. By conducting molecular dynamics (MD) simulation using this model, we first studied basic properties of the rotational motion of F_O in equilibrium. We then imposed non-equilibrium condition on F_O. Realistic non-equilibrium condition is, of course, to apply an electrochemical potential across the membrane to drive the proton current. Actually, we employed such a non-equilibrium condition to investigate possible proton pathway in F_O, and its result will be presented elsewhere.[4] Here we consider another non-equilibrium condition where the c-ring and the a-subunit are in contact with different heat baths with different temperatures, respectively. This non-equilibrium condition is reminiscent of the Feynman ratchet,[5] which has been proposed as a possible mechanism of the actomyosin molecular motor.[6]

2. Model and Method

2.1. *3D structure of F_O*

Since the whole 3D structure of the F_O portion has not yet been determined, especially of the a-subunit, we used the 3D structure of the a_1c_{12} complex (PDB-ID:1C17[7]) which has been modeled using the structures of the transmembrane helices (TMHs) determined by NMR together with the disulfide cross-linking data. We modified the c-ring structure to fulfill the exact rotational symmetry. We did not explicitly include the b-subunit, which is a constituent of F_O, although the role of the b-subunit in proton translocation and c-ring rotation is under debate.[8,9] We instead take into account the role of the b-subunit implicitly as a spring applied to TMH2[10] to keep the a-subunit from freely rotating and from moving away from c-ring

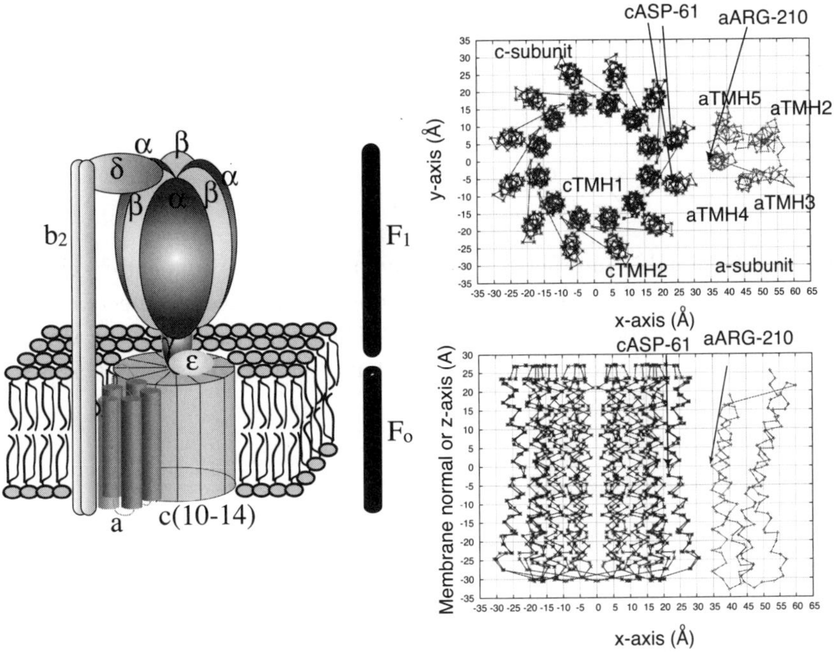

Fig. 1. Left figure: Schematic view of the F_1F_O-ATP synthase. The solvent-exposed F_1 unit (top) consists of subunits $\alpha_3\beta_3\gamma\delta\epsilon$ and the membrane-embedded F_O unit (bottom) consists of subunits ab_2c_{10-14}. Right figures: Initial structure of F_O-complex were modified from the model structure (PDB code:1c17[7]). Positions of all amino acids (C_α) are represented as points.

2.2. *Protein model and interactions*

A coarse-grained (CG) model, in which protein molecules are represented by α-carbon chain and the native contact between $C_\alpha$atoms are kept by Hookean potential, has been shown to properly reproduce both isotropic and anisotropic thermal fluctuations.[11] This CG model is often referred to as the elastic network model and is employed for the c-ring and the a-subunit, respectively. The spring constant k for the "bond" between the i-th and j-th C_α atoms was set at $k = 6$ ($|j - i| = 1$), $k = 3$ ($|j - i| = 2$), and $k = 0.6$ ($|j - i| \geq 3$) to better reproduce the eigenvalue spectrum of principal components of thermal fluctuations calculated from a more realistic all-atom model MD simulation.[11] In this study, a C_α atom pair ($|j-i| \geq 3$) the distance of which is less than 10 Å in the native (i.e., PDB) structure was considered to form a native contact. For the details of the elastic network model, see Ref. 12 and references therein.

We then introduce the interactions between the a-subunit and the c-ring. We followed the interaction model used in our previous study of actomyosin molecular motor, where two inter-molecular interactions, the Coulombic and Lennard-Jones-type van der Waals interactions, were considered:[12]

$$V_{\text{inter}} = \sum_{i,j} \frac{e^2}{4\pi\epsilon_0\epsilon_r} \frac{q_i q_j}{r_{ij}} + \sum_{i,j} 4\varepsilon \left[\left(\frac{\sigma}{r_{ij}} \right)^{12} - \left(\frac{\sigma}{r_{ij}} \right)^{6} \right]. \tag{1}$$

For the Coulombic interactions, negative and positive elementary charges are assigned for the acidic and basic amino acid residues, respectively. We set $\epsilon_r = 4$, $\varepsilon \approx 0.05$ kcal/mol, $\sigma = 7$ Å. The cutoff length for the van der Waals interaction and that for the Coulomb interaction were set at 10 Å and 20 Å, respectively.

2.3. *Molecular Dynamics*

The Langevin equation,

$$m\frac{d^2\mathbf{r_i}}{dt^2} = -\nabla_i U - m\zeta\frac{d\mathbf{r_i}}{dt} + \xi_i(t), \tag{2}$$

was employed for the time evolution of the F_O model system as defined above. U is the sum of the elastic energy and the interaction energy between the a-subunit and the c-ring, $\mathbf{r_i}$ and m represent the position and mass of atom i, respectively. $m\zeta$ is the friction coefficient and $\xi_i(t)$ is the Gaussian white noise, the correlation function of which is $\langle\xi(t)\xi(0)\rangle = 2\zeta k_B T\delta(t)\delta_{ij}$. Unit time in the MD simulation using this CG model approximately corresponds to 0.3 ps (unit mass corresponding to 100 g/mol), and we used $\zeta = 0.01$. The temperature was set so as to reproduce the root mean square fluctuations (RMSF) of C_α atoms calculated from an all-atom model MD simulation at room temperature ($k_B T = 0.6$ in this study)

3. Rotational Motion of the c-ring in Equilibrium

We first study basic properties of the rotational motion of F_O in equilibrium (in other words, a study on the tribology of F_O). To study the rotational motion of the c-ring, the rotational angle of c-ring relative to the a-subunit was monitored. The rotational center of the c-ring is the origin in the coordinate system shown in Fig. 1. The rotational angle is defined as the azimuthal angle of the specified residue (cAsp61) in the c-ring. Figure 2a shows the time evolution of the rotational angle in the absence/presence of the a-subunit. In the absence of the a-subunit, c-ring shows free rotational

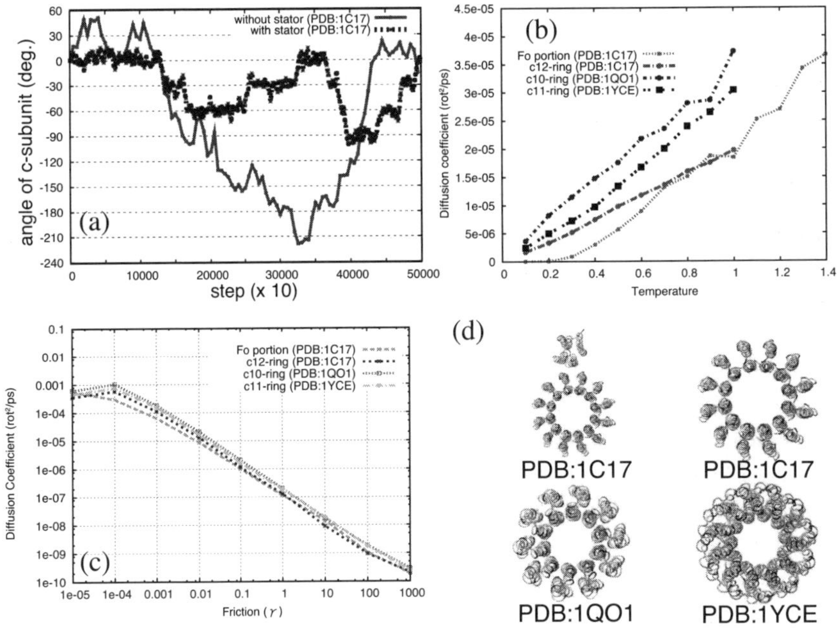

Fig. 2. (a) Trajectories of the c-ring rotation with and without the a-subunit. (b) Temperature dependence of the rotational diffusion constant: three different c-rings consisting of different number of c-subunits were studied for comparison (c_{12}, c_{10}, c_{11} rings are modeled from PDB structures, 1C17[7] , 1QO1,[13] 1YCE,[14] respectively; see (d)). (c) ζ dependence of the rotational diffusion coefficient. (d) PDB structures studied here viewed from the cytoplasmic side. Total number of the residues in the c-ring is 948, 790, and 979 for c_{12}, c_{10}, and c_{11}, respectively.

diffusion, as is expected. In the presence of the a-subunit, c-ring shows stepwise rotational motion, the step angle corresponding to one c-subunit. We then calculate the rotational diffusion coefficient from the mean square displacement (MSD) of the angle, i.e., $2Dt = \langle(\theta(t) - \theta(0))^2\rangle$. Figures 2b and 2c show the temperature and the ζ dependence of diffusion coefficients. Basically, the Einstein's relation $D = k_BT/\gamma$ (where $\gamma = m\zeta$ in this study) suffices except for the following two points. One point is the roll-over seen in Fig. 2c in the small ζ region ($\zeta < 0.0001$) . This is simply due to the fact that the ballistic motion dominates over the diffusional motion within the timescale where MSD was calculated. The other is the suppressed diffusion seen in Fig. 2b in the presence of the a-subunit at temperatures lower than $T \approx 0.3$. This transition-like behavior is noteworthy and reminds us of the glass-transition, as often observed in the dynamics of protein molecules.[15]

Interestingly, $T \approx 0.3$ corresponds to 150 K, which coincides closely with the glass transition temperature of proteins. This result also suggests that internal flexibility (i.e., thermal fluctuation) of F_O plays an important role in lowering the rotational friction.

4. Non-equilibrium Rotational Motion

In living cells, F_O is under non-equilibrium condition in that there is an electrochemical potential gradient across the cell membrane and the steady proton current is driven across the membrane. By imposing this type of non-equilibrium condition, we studied the proton pathway in F_O by using the same CG model as employed in this study. We found a couple of interesting results, which include a bottleneck formed by six positively charged amino-acids located in the middle of the putative central channel formed by four TMHs (TMH2, 3, 4, and 5) in the a-subunit, and a proton reservoir located in the periplasmic side of the putative central channel which is formed by conserved acidic residues.[4]

Imagine that protons are driven to pass through the channel in the a-subunit, then the protons would do work against the a-subunit, causing non-equilibrium fluctuations in the a-subunit (and perhaps in the c-subunit as well). In the extreme case, the proton current might cause temperature graduation in F_O. We here present a preliminary result of non-equilibrium MD simulation where the c-ring and the a-subunit are in contact with different heat baths of different temperatures, T_2 and T_1, respectively. This non-equilibrium situation of F_O can be regarded as a molecular incarnation of the Feynman ratchet.[5]

Figure 3 shows the rotation motions of the c-ring observed in MD simulation under the three different conditions. First, in equilibrium, we can only see free rotational diffusion without directionality Figs. 3b and 3d, as already mentioned. When $T_1 > T_2$ ($T_1 = 2 \times T_2$, where T_2 corresponds to the room temperature), the net directionality arises (Figs. 3a and 3d). The direction of the observed rotational motion is the ATP synthesis direction. This is interesting because in the ATP synthesis mode, the proton current empowered by the electrochemical potential across the membrane would do work against the a-subunit first (note that the entrance of the proton channel is found in the a-subunit[4]), and hence they might cause more significant non-equilibrium fluctuation to the a-subunit than the c-ring. When $T_2 > T_1$ ($T_1 = 2 \times T_1$, where T_1 corresponds to the room temperature), the net direction is observed again, and, moreover, the direction is reversed. These observations are consistent with the way the Feynman ratchet works.

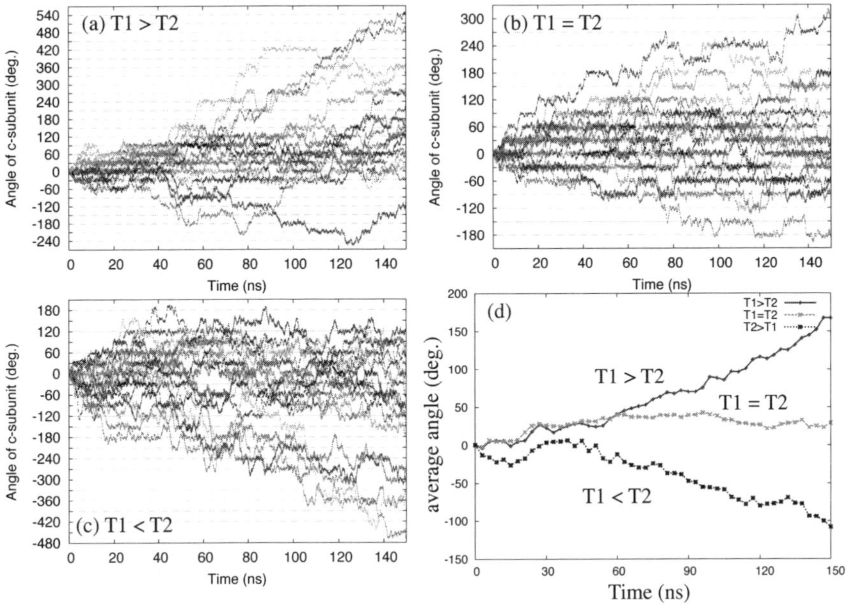

Fig. 3. (a) 20 angle trajectories of c-ring for $T_1 > T_2$, (b) those for $T_1 = T_2$, and (c) those for $T_1 < T_2$. T_1 and T_2 are the temperatures of the heat baths, with which the a-subunit and the c-ring are in contact, respectively. (d)Average trajectory of 20 runs for three different conditions.

5. Discussion and Conclusion

We simulated rotational diffusion of the c-ring in the F_O rotary motor, based on present available 3D structure under the equilibrium and non-equilibrium conditions. From the equilibrium simulation, we found how the protein softness affects the friction of the rotational motion. From the non-equilibrium simulation, we found a Feynman-ratchet-like mechanism inherent in the F_O. In both results, the softness of protein structure plays an important role. As for the former result, the role of the protein softness is rather obvious. As for the latter result, protein softness may be involved in maintaining the local high temperature (or high energy state) which could be caused by the proton current. There is, of course, much room for further study, and we have to carefully examine to what extent the coarse grained models as used in this study could yield faithful results.

References

1. W. Junge, H. Lill and S. Engelbrecht, Trends. Biochem. Sci. **22** (1997) 123.

2. T. Elston, H. Wang and G. Oster, Nature **391** (1998) 510.
3. A. Aksimentiev, I. A. Balabin, R. H. Fillingame and K. Schulten, Biophys. J. **86** (2004) 1332.
4. H. Yamasaki and M. Takano, manuscript in preparation.
5. R. P. Feynman, R. B. Leighton and M. Sands, *The Feynman Lectures on Physics* (Addison-Wesley, Reading, MA, 1966), vol.1, chap. 46.
6. R. D. Vale and F. Oosawa, Adv. Biophys. **26** (1990) 97.
7. V. K. Rastogi and M. E. Girvin, Nature **402** (1999) 263.
8. S. D. Dunn, M. Revington, D. J. Cipriano and B. H. Shilton, J. Bioenerg. Biomem. **32** (2000) 347.
9. S. Ono, N. Sone, M. Yoshida and T. Suzuki, J. Biol. Chem. **279** (2004) 33409.
10. S. B. Vik and R. R. Ishmukhametov, J. Bioener. Biomem. **37** (2005) 445.
11. M. Takano, J. Higo, H. K. Nakamura and M. Sasai, Nat. Comput. **3** (2004) 377.
12. M. Takano, T. P. Terada and M. Sasai, manuscript in preparation.
13. D. Stock, A. G. W. Leslie and J. E. Walker, Science **286** (1999) 1700.
14. T. Meier, P. Polzer, K. Diederichs, W. Welte and P. Dimroth, SCIENCE **308** (2005) 659.
15. R. H. Austin, L. Einstein, H. Frauenfelder and I. C. Gunsalns, Biochemistry **14** (1975) 5355.

PART B

Nonequilibrium Statistical Physics and Related Topics

THERMODYNAMIC TIME ASYMMETRY AND NONEQUILIBRIUM STATISTICAL MECHANICS

PIERRE GASPARD

Center for Nonlinear Phenomena and Complex Systems,
Université Libre de Bruxelles, Code Postal 231, Campus Plaine,
B-1050 Brussels, Belgium

A review of recent advances in nonequilibrium statistical mechanics is presented. These new results explain how the time reversal symmetry is broken in the statistical description of nonequilibrium processes and the thermodynamic entropy production finds its origin in the time asymmetry of nonequilibrium fluctuations. These advances are based on new relationships expressing the time asymmetry in terms of the temporal disorder and the probability distributions of nonequilibrium fluctuations. These relationships apply to driven Brownian motion and molecular motors.

Keywords: Microreversibility; entropy production; relaxation; diffusion; out-of-equilibrium nanosystems.

1. Introduction

Nonequilibrium statistical mechanics is undergoing tremendous advances with the discovery of universal relationships ruling molecular fluctuations near and far from equilibrium. These new relationships have been discovered thanks to dynamical systems theory where new concepts have been introduced to understand chaotic behavior. In this context, the issue of the initial conditions of a dynamical system has been discussed about the phenomenon of sensitivity to initial conditions. The focus on this issue have disclosed striking features on the role of initial conditions in the time asymmetry of nonequilibrium processes.

It is our purpose to give an overview of these recent results and their applications to out-of-equilibrium nanosystems. Indeed, at the nanoscale, the fluctuations can no longer be neglected as it is the case at the macroscale. Accordingly, the processes taking place in nanosystems should be described in terms of the probability distributions of molecular fluctuations. The fact is that many nanosystems find their relevance because they function under nonequilibrium conditions, as for molecular motors or electronic conducting nanodevices. The new advances in nonequilibrium statistical mechanics thus establish the bases for a statistical thermodynamics of out-of-equilibrium nanosystems. Remarkably, these recent developments provide an unprecedented insight into our understanding of biological phenomena from the viewpoint of statistical thermodynamics. This insight extends much beyond the already important results obtained with the macroscopic thermodynamics of irreversible processes,[1] which did not deal with the fluctuations and their temporal properties. By carrying out the connection between the macroscale and the molecular aspects at the nanoscale, nonequilibrium statistical mechanics is nowadays able to bridge the gaps between molecular biology, bioenergetics, and the thermodynamic laws thanks to the advent of the new results about nonequilibrium fluctuations.

The review is organized as follows. The breaking of time reversal symmetry in nonequilibrium statistical mechanics is discussed in Sec. 2 and illustrated in Sec. 3 with the relaxation modes of diffusion. We explain in Sec. 4 how the thermodynamic arrow of time finds its origin in the time asymmetry of temporal disorder in nonequilibrium fluctuations. In Sec. 5, we give an overview of the so-called fluctuation theorems. The case of molecular motors is presented in Sec. 6. Conclusions and perspectives are given in Sec. 7.

2. The Breaking of Time Reversal Symmetry in Nonequilibrium Statistical Mechanics

The second law of thermodynamics is expressed in terms of the concept of entropy. This quantity was interpreted by Boltzmann as a characterization of the disorder in the probability distribution of the positions and velocities of the particles composing the system. Later, Gibbs pointed out that the microscopic definition of entropy requires coarse graining in one way or another. There exist different versions of coarse graining. The first was introduced by Boltzmann who obtained his famous H-theorem by describing dilute gases in terms of one-particle distributions. In this way, he neglected the statistical correlations between several particles, which may be justified for dilute gases but allows the corresponding entropy to vary in time. Another version consists in tracing out the degrees of freedom of the environment of a subsystem, as done for a Brownian particle. *In fine*, the phase space can be partitioned into grains or cells to get the occupation probabilities of these cells, as suggested by Gibbs in 1902. Since the work of Sadi Carnot, coarse graining is inherently associated with the idea of entropy because the efficiency of steam engines is a property of the steam microscopic degrees of freedom manipulated by some macroscopic piston on a very coarse-grained level. Only the coarse-grained entropy may evolve in time as the consequence of the dynamical property of mixing, also introduced by Gibbs. Indeed, the fine-grained entropy remains constant in time.

However, the breaking of the time reversal symmetry can be formulated in nonequilibrium statistical mechanics at a more fundamental level without referring from the start to the thermodynamic entropy, but instead discussing the symmetry of the probability distribution of the positions $\mathbf{r}$ and velocities $\mathbf{v}$ of the particles in classical mechanics:

$$p(\mathbf{\Gamma}; t) = p(\mathbf{r}, \mathbf{v}; t) \tag{1}$$

or the matrix density in quantum mechanics. These quantities obey time evolution equations that is the Liouville equation for classical systems:

$$\partial_t \, p = \{H, p\}_{\text{Poisson}} \equiv \hat{L} \, p \tag{2}$$

where $\{H, \cdot\}_{\text{Poisson}}$ denotes the Poisson bracket with the Hamiltonian function H, or the Landau-von Neumann equation for quantum systems. Actually, the time evolution of the probability distribution is induced by the deterministic equations of motion which are Newton's or Hamilton's equations of classical mechanics and Schrödinger's equation of quantum mechanics. These equations are deterministic in the sense that their solutions are uniquely given in terms of their initial conditions according to Cauchy's theorem. In classical systems, the initial condition is a point $\mathbf{\Gamma} = (\mathbf{r}, \mathbf{v})$ taken in the phase space of the positions and velocities of the particles while the initial condition is a wavefunction taken in the Hilbert space of a quantum system.

Probability distributions are introduced because the initial state can never be prepared with an infinite precision on uncountable continua such as phase spaces or Hilbert spaces. On such continua, there is always a dichotomy between the existence of a point-like initial condition and the necessarily unprecise knowledge of this latter. As a result, the predictability of the trajectory becomes a concern in systems with sensitivity to initial conditions – the so-called chaotic systems. Dynamical chaos guarantees Gibbs' mixing property, which allows us to understand relaxation processes in many systems. It is also important to justify the use of a statistical description in terms of probability distributions beyond the Lyapunov time characteristic of the sensitivity to initial conditions.[2]

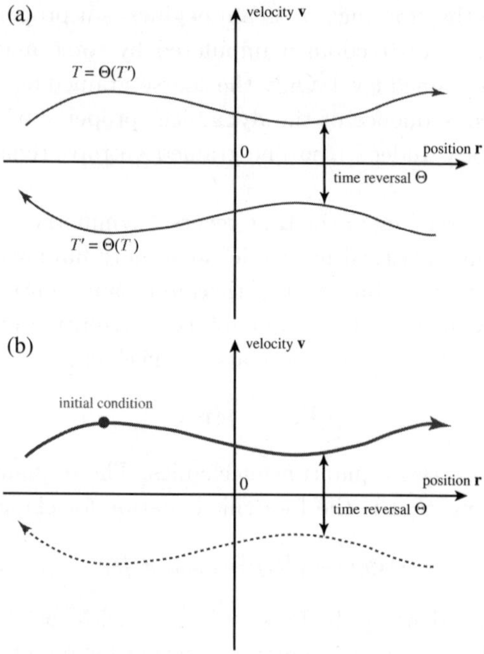

Fig. 1. (a) If Newton's equation is time reversal symmetric, the time reversal $\Theta(T)$ of every solution T is also a solution, as here shown in the phase space of the positions and velocities of the system. (b) If the time reversal $\Theta(T)$ is physically distinct from the trajectory T, the selection of the trajectory T by the initial condition gives a unit probability to T and zero to all the other solutions including the time reversal image $\Theta(T)$, therefore breaking the time reversal symmetry.

It turns out that the breaking of time reversal symmetry is another issue concerning the initial conditions in chaotic as well as non-chaotic systems, as shown by the following discussion carried out in the framework of classical mechanics. It is well known that Newton's equation is symmetric under time reversal:

$$\Theta(\mathbf{r}, \mathbf{v}) = (\mathbf{r}, -\mathbf{v}) \tag{3}$$

which leaves the positions $\mathbf{r}$ unchanged and reverses the velocities $\mathbf{v} = d\mathbf{r}/dt$, if the Hamiltonian is an even function of the velocities. The symmetry – called microreversibility – means that the time reversal of a solution of Newton's equation is also a solution of this equation (see Fig. 1a). Each initial condition selects a precise solution of Newton's equations, which depicts a trajectory $T = \{\mathbf{\Gamma}(t)|t \in \mathbb{R}\}$ in the phase space $\mathbf{\Gamma} = (\mathbf{r}, \mathbf{v})$. We may say that a trajectory T is *physically distinct* from another trajectory

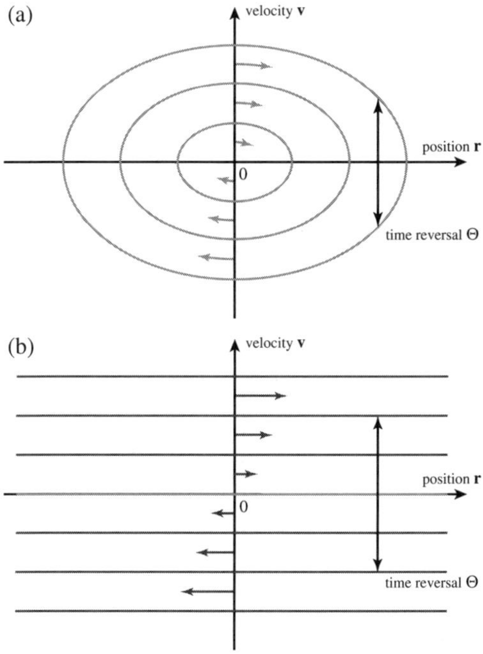

Fig. 2. (a) Phase portrait of the harmonic oscillator (4). All its trajectories are ellipses which are self-reversed. Consequently, the selection of an initial condition does not break the time reversal symmetry in this system. (b) Phase portrait of the free particle (5). Here, the trajectories are physically distinct from their image under time reversal, except if the velocity vanishes. Consequently, the selection of an initial condition (with a non-vanishing velocity) breaks the time reversal symmetry in this system.

if they do not coincide in the phase space of the system.

If the trajectory T followed by the system during its time evolution is physically distinct from its time reversal, $\Theta(T) \neq T$, it turns out that the time reversal symmetry is broken in the system. This is the same as for other symmetry breaking phenomena in condensed matter physics. For instance, the double-well potential $V(x) = (x^2/2) - (x^4/4)$ is symmetric under parity $x \to -x$. This symmetry is broken if the system is found in one of the wells, either $x = +1$ or $x = -1$. This illustrates the general result that the solution of an equation may have a lower symmetry than the equation itself. This phenomenon of symmetry breaking applies to time reversal as well. Although Newton's equation is time reversal symmetric, its solutions do not necessarily have the symmetry. Therefore, the selection of a trajectory by the initial condition can break the time reversal symmetry.

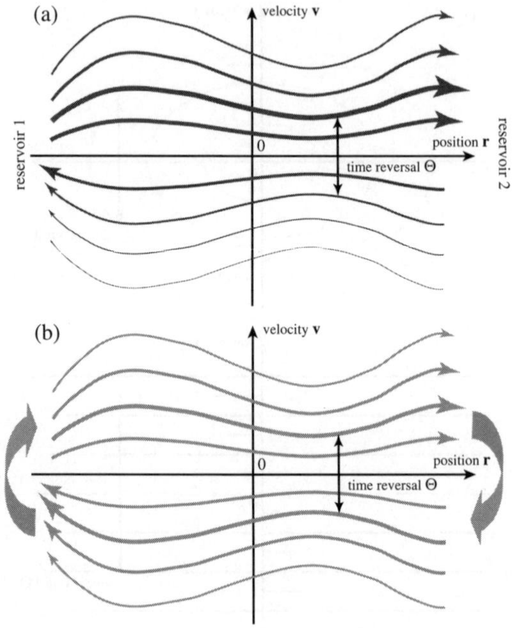

Fig. 3. (a) Phase portrait of a nonequilibrium system with particles diffusing between the high concentration reservoir 1 and the low concentration reservoir 2. The trajectories from 1 to 2 typically have a higher probability weight than the trajectories from 2 to 1 in order that the mean current flows from 1 to 2. The probability weight is indicated by the thickness of the trajectories, showing the time asymmetry of the probability distribution. (b) Phase portrait at equilibrium if the contacts with the reservoirs are closed, in which case detailed balance is satisfied and the probability distribution is time reversal symmetric.

This happens if the trajectory selected by the initial condition is physically distinct from its time reversal (see Fig. 1b).

This time reversal symmetry breaking does not manifest itself in every system. In particular, all the trajectories of Newton's equation:

$$m\frac{d^2r}{dt^2} = -kr \tag{4}$$

of the harmonic oscillator are the ellipses $E = (mv^2 + kr^2)/2$ in the phase space (r, v). Each ellipsis is mapped onto itself by time reversal $\Theta(r, v) = (r, -v)$. All the trajectories are thus self-reversed in the harmonic oscillator (see Fig. 2a). In contrast, almost all the trajectories are distinct from their time reversal in the case of the free particle of Newton's equation:

$$m\frac{d^2r}{dt^2} = 0 \tag{5}$$

Indeed, the trajectories are the straight lines $r(t) = r(0)+v(0)t$, $v(t) = v(0)$, which are distinct from their reversal if the velocity is non vanishing, $v(0) \neq 0$ (see Fig. 2b). Therefore, the selection of an initial condition may already break the time reversal symmetry in this simple system. *A fortiori*, this breaking can also happen in a chaotic system with a spectrum of positive Lyapunov exponents indicating many stable and unstable directions in phase space. These directions are mapped onto each other but physically distinct so that the time reversal symmetry will be broken if one specific direction is selected by the initial condition.[3,4]

In statistical mechanics, each initial condition – and thus each phase-space trajectory – is weighted with a probability giving its statistical frequency of occurrence in a sequence of repeated experiments. This is in particular the case for a nonequilibrium system with particles diffusing between two reservoirs at different concentrations. After some transients, the system reaches a nonequilibrium steady state which can be described by an invariant probability distribution. Averaging over this distribution gives a mean current of diffusing particles from high to low concentrations. In this case, the phase-space trajectories issued from the high concentration reservoir typically have a higher probability weight than the trajectories from the low concentration reservoir (see Fig. 3a). Since the set of these latter trajectories contain the time reversal of the former ones, we conclude that the time reversal symmetry is broken by the nonequilibrium invariant probability distribution:

$$p_{\rm neq}(\Theta\mathbf{\Gamma}) \neq p_{\rm neq}(\mathbf{\Gamma}) \tag{6}$$

Of course, if the contacts with the reservoirs are closed, the invariant probability becomes the equilibrium one after relaxation, in which case detailed balance is satisfied and the time reversal symmetry is restored:

$$p_{\rm eq}(\Theta\mathbf{\Gamma}) = p_{\rm eq}(\mathbf{\Gamma}) \tag{7}$$

(see Fig. 3b). Since the nonequilibrium invariant probability distributions are stationary solutions of Liouville's equation (2), we here have a similar symmetry breaking phenomenon as for Newton's equation. The nonequilibrium stationary density (6) is a solution of Liouville's equation with a lower symmetry than the equation itself. In this way, irreversible behavior can be described by weighting differently the trajectories T and their time reversal images $\Theta(T)$ with a probability measure.

3. The Relaxation Modes of Diffusion

In this section, we show in specific systems that the breaking of time reversal symmetry is indeed the fact of systems in nonequilibrium states. We consider Hamiltonian systems with chaotic diffusion such as the multibaker map,[5–7] the hard-disk Lorentz gas, and the Yukawa-potential Lorentz gas.[8] The Newtonian dynamics of these systems is Hamiltonian and time reversal symmetric. These systems sustain the transport property of diffusion because of their spatial extension. Moreover, the dynamics is periodic in space as for the motion of electrons or impurities in a crystal.

If the initial conditions are taken out of equilibrium, the probability distribution undergoes a transient relaxation toward a uniform state corresponding to the thermodynamic equilibrium. This relaxation can be decomposed into modes which are special solutions of Liouville's equation (2):

$$p(\mathbf{r}, \mathbf{v}; t) = C \exp(s_{\mathbf{k}} t) \, \Psi_{\mathbf{k}}(\mathbf{r}, \mathbf{v}) \tag{8}$$

These solutions are spatially periodic $\Psi_{\mathbf{k}}(\mathbf{r}, \mathbf{v}) \sim \exp(i\mathbf{k} \cdot \mathbf{r})$ with a wavelength $\lambda = 2\pi/k$ typically much longer than the periodicity of the crystal. The solutions (8) are exponentially damped in time at the rate

$$-s_{\mathbf{k}} = \mathcal{D}\mathbf{k}^2 + O(\mathbf{k}^4) \tag{9}$$

vanishing quadratically with the wavenumber $\mathbf{k}$, which defines the diffusion coefficient given by the Green-Kubo formula:

$$\mathcal{D} = \int_0^\infty \langle v_x(0) v_x(t) \rangle_{\text{eq}} \, dt \tag{10}$$

It turns out that the relaxation modes (8) can be constructed in phase space as generalized eigenstates of the Liouvillian operator of Eq. (2):

$$\hat{L} \, \Psi_{\mathbf{k}} = s_{\mathbf{k}} \Psi_{\mathbf{k}} \tag{11}$$

with the associated eigenvalue $s_{\mathbf{k}}$ given by a so-called Pollicott-Ruelle resonance.[2] The generalized eigenstates $\Psi_{\mathbf{k}}$ do not exist as functions but as distributions of Schwartz type, which are defined on some functional space of test functions. The remarkable feature of the construction is that it can be carried out without a specific coarse graining since the distribution $\Psi_{\mathbf{k}}$ is defined on a whole functional space of possible test functions. The use of such test functions may be considered as coarse graining but the generalized eigenstate does not depend on the choice of a specific test function taken in the functional space.

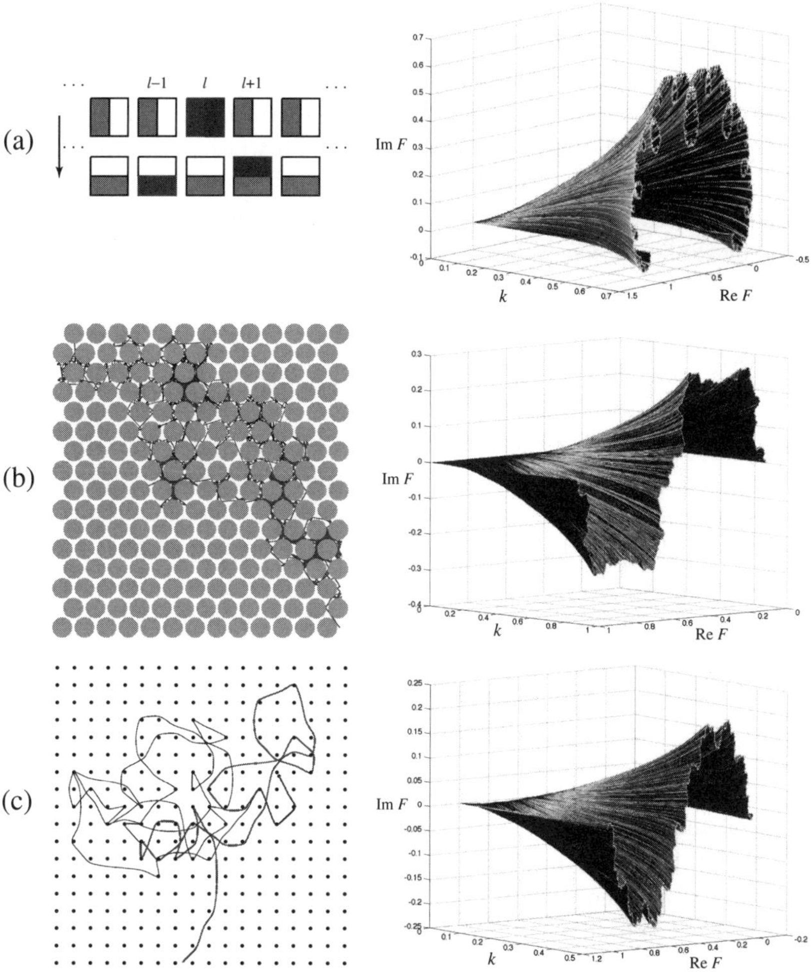

Fig. 4. The relaxation modes of diffusion in (a) the multibaker map,[5–7] (b) the hard-disk Lorentz gas, and (c) the Yukawa-potential Lorentz gas.[8] The left-hand column shows the mechanism of diffusion of particles in these systems. In the right-hand column, the cumulative function (12) is depicted in the complex plane $(\mathrm{Re}\, F, \mathrm{Im}\, F)$ versus the wavenumber k, for each system. If the wavenumber vanishes $k = 0$, the cumulative function reduces to the straight line $\mathrm{Im}\, F = 0$ between the points $\mathrm{Re}\, F = 0$ and $\mathrm{Re}\, F = 1$, which represents the microcanonical equilibrium state.

The densities $\Psi_{\mathbf{k}}(\mathbf{r},\mathbf{v})$ are singular along the stable manifolds W_s of phase space, but nevertheless smooth along the unstable manifolds W_u. Since the stable and unstable manifolds are mapped onto each other by time reversal $W_u = \Theta(W_s)$, but are physically distinct $W_u \neq W_s$, the densities $\Psi_{\mathbf{k}}(\mathbf{r},\mathbf{v})$ are solutions of Liouville's equation which break the time reversal symmetry, as expected for solutions corresponding to unidirectional exponential decay. The time asymmetry of the relaxation modes can be displayed by explicitly constructing the eigenstates thanks to their cumulative function

$$F_{\mathbf{k}}(\theta) = \int_0^{\theta} \Psi_{\mathbf{k}}(\mathbf{r}_{\theta'},\mathbf{v}_{\theta'})\, d\theta' = \lim_{t\to\infty} \frac{\int_0^{\theta} \exp\left[i\mathbf{k}\cdot(\mathbf{r}_t - \mathbf{r}_0)_{\theta'}\right] d\theta'}{\int_0^{2\pi} \exp\left[i\mathbf{k}\cdot(\mathbf{r}_t - \mathbf{r}_0)_{\theta'}\right] d\theta'} \tag{12}$$

obtained by integrating their density $\Psi_{\mathbf{k}}(\mathbf{r},\mathbf{v})$ over some line $(\mathbf{r}_\theta,\mathbf{v}_\theta)$ in phase space.[8] Because of the singular character of the density in the stable directions, these cumulative functions depict fractal curves in the complex plane $(\mathrm{Re}\, F_{\mathbf{k}}, \mathrm{Im}\, F_{\mathbf{k}})$. These fractal curves are depicted in Fig. 4 for the multibaker map[5–7] and the aforementioned Lorentz gases.[8]

The fractal dimension D_{H} of these curves is given by the root of the equation:[8]

$$\mathcal{D}\mathbf{k}^2 \simeq -\mathrm{Re}\, s_{\mathbf{k}} = \lambda(D_{\mathrm{H}}) - \frac{h(D_{\mathrm{H}})}{D_{\mathrm{H}}} \tag{13}$$

where $\mathcal{D}$ is the diffusion coefficient while λ and h are the positive Lyapunov exponent and the Kolmogorov-Sinai entropy per unit time defined at the value D_{H} of the fractal dimension. Given that the Kolmogorov-Sinai entropy tends to the Lyapunov exponent as the fractal dimension approaches unity, this latter becomes

$$D_{\mathrm{H}} = 1 + \frac{\mathcal{D}}{\lambda}\mathbf{k}^2 + O(\mathbf{k}^4) \tag{14}$$

for small values of the wavenumber.[8] Since the relaxation modes are singular in the stable directions but smooth in the unstable ones, the fractal character of their cumulative function is the direct manifestation of the breaking of the time reserval symmetry by these modes.

We notice that Eq. (13) is an extension of the escape-rate formulae giving each transport coefficient in terms of the Lyapunov exponents and the Kolmogorov-Sinai entropy per unit time on a fractal repeller.[9,10] Already in this framework, the escape rate is associated with nonequilibrium decaying states which break the time reversal symmetry.

The density of the nonequilibrium steady state of gradient $\mathbf{g}$ can be obtained from the densities (8) of the relaxation modes according to

$$\Psi_{\mathbf{g}}(\mathbf{\Gamma}) = -i\mathbf{g} \cdot \frac{\partial}{\partial \mathbf{k}} \Psi_{\mathbf{k}}(\mathbf{\Gamma})\Big|_{\mathbf{k}=0} = \mathbf{g} \cdot \left[\mathbf{r}(\mathbf{\Gamma}) + \int_0^{-\infty} \mathbf{v}(\mathbf{\Phi}^t \mathbf{\Gamma})\, dt\right] \tag{15}$$

where $\mathbf{\Phi}^t$ denotes the Hamiltonian flow.[6] Because of Eq. (15), the nonequilibrium steady states have similar singularities as the relaxation modes. Their distribution is smooth along the unstable manifolds but singular in the stable directions. Their cumulative function is the nondifferentiable Takagi function in the multibaker map and its generalizations in the Lorentz gases.[6] This singular character is thus the manifestation of the breaking of the time reversal symmetry by the invariant probability distribution of nonequilibrium steady states.

4. Experimental Evidence of the Time Asymmetry of Nonequilibrium Fluctuations

The breaking of the time reversal symmetry by the invariant probability distribution describing the nonequilibrium steady states has direct experimental consequences, which have been evidenced in recent experiments in the group of Professor Sergio Ciliberto at the ENS of Lyon (France) on the driven Brownian motion of a micrometric particle trapped by an optical tweezer in a moving fluid and the Nyquist thermal noise of a *RC* electric circuit driven out of equilibrium by a current.[11,12]

In driven Brownian motion, the position of the particle can be monitored with nanometric resolution thanks to an interferometer. Long time series of the position of the Brownian particle have been recorded in the frame of the optical trap for a driving by the surrounding fluid with the speeds u and $-u$. Examples of such paths are depicted in Fig. 5a. Figure 5b gives the corresponding stationary probability distributions of the position z, showing the effect of the drag due to the moving fluid. The observations are well described by an overdamped Langevin equation including the force exerted by the potential of the laser trap, the drag force of the moving fluid, the viscous firction force, and the Langevin force of the thermal fluctuations.[11,12] The long time series allow us to measure the probabilities of paths $\boldsymbol{\omega} = \omega_0\omega_1\omega_2\ldots\omega_{n-1}$ of varying resolution ε on the position and sampled every time interval τ. We can compare the probability of a path in the process of speed u with the probability of the time-reversed path $\boldsymbol{\omega}^{\mathrm{R}} = \omega_{n-1}\ldots\omega_2\omega_1\omega_0$ in the process of opposite speed $-u$. As for the trajectories, a time-reversed path is typically distinct from the corresponding path, $\boldsymbol{\omega}^{\rm R} \neq \boldsymbol{\omega}$. The coincidence only happens for the rare self-reversed paths. According to detailed balance, we expect that the probabilities of the paths and their time reversal are equal if $u = 0$:

$$\text{equilibrium:} \qquad P_0(\omega_0\omega_1 \ldots \omega_{n-1}) = P_0(\omega_{n-1} \ldots \omega_1\omega_0) \tag{16}$$

However, this equality is not expected if $u \neq 0$:

$$\text{out of equilibrium:} \qquad P_u(\omega_0\omega_1 \ldots \omega_{n-1}) \neq P_{-u}(\omega_{n-1} \ldots \omega_1\omega_0) \tag{17}$$

This difference also affects the decay of these path probabilities. Their mean decay rates characterize the temporal disorder or dynamical randomness and are called the (ε, τ)-entropy per unit time:[13]

$$h = \lim_{n\to\infty} -\frac{1}{n\tau} \sum_{\omega_0\omega_1\ldots\omega_{n-1}} P_u(\omega_0\omega_1 \ldots \omega_{n-1}) \ln P_u(\omega_0\omega_1 \ldots \omega_{n-1}) \tag{18}$$

and time-reversed (ε, τ)-entropy per unit time:[14]

$$h^{\rm R} = \lim_{n\to\infty} -\frac{1}{n\tau} \sum_{\omega_0\omega_1\ldots\omega_{n-1}} P_u(\omega_0\omega_1 \ldots \omega_{n-1}) \ln P_{-u}(\omega_{n-1} \ldots \omega_1\omega_0) \tag{19}$$

The supremum of the dynamical entropy (18) over all the possible coarse grainings defines the famous Kolmogorov-Sinai entropy per unit time, which equals the sum of positive Lyapunov exponents in closed dynamical systems according to Pesin's theorem.[2] Instead, the time-reversed entropy per unit time (19) has been recently introduced motivated by the difference (17) expected in nonequilibrium processes.[14] Contrary to the thermodynamic entropy which measures the disorder of the probability distribution in the phase space at a given time, the entropies per unit time characterize the disorder displayed by the process along the time axis, as in the successive pictures of a movie. We thus speak of temporal disorder or dynamical randomness for the property characterized by the quantities (18) and (19). To avoid a possible confusion with the standard thermodynamic entropy, we shall respectively call them the forward and reversed randomnesses in the following.

The most remarkable result is that the difference between these randomnesses gives the thermodynamic entropy production according to

$$\frac{1}{k_{\rm B}}\frac{d_{\rm i}S}{dt} = \lim_{\varepsilon,\tau\to 0} \left[h^{\rm R}(\varepsilon,\tau) - h(\varepsilon,\tau)\right] \geq 0 \tag{20}$$

where $k_{\rm B}$ is Boltzmann's constant.[14] This difference is the Kullback-Leibler distance between the path probabilities $P_u(\boldsymbol{\omega})$ and $P_{-u}(\boldsymbol{\omega}^{\rm R})$, also known under the name of relative entropy, and is therefore always non-negative in agreement with the second law of thermodynamics. The formula (20) is remarkable because it connects the arrow of time of macroscopic thermodynamics to another arrow of time observed at the mesoscopic scales in the temporal disorder of the nonequilibrium fluctuations. The time asymmetry of this temporal disorder manifests itself in the difference $h^{\mathrm{R}} - h$. The formula (20) thus shows that the thermodynamic arrow of time finds its origin in the time asymmetry of temporal disorder in the nonequilibrium fluctuations.

The formula (20) is also remarkable because it provides a rational foundation to the intuitive idea that dynamical order arises in a system driven out of equilibrium, as expressed by the

Theorem of nonequilibrium temporal ordering:[15] *In nonequilibrium steady states, the typical paths are more ordered in time than their corresponding time reversals in the sense that their temporal disorder characterized by h is smaller than the temporal disorder of the corresponding time-reversed paths characterized by h^{R}.*

This theorem expresses the fact that the molecular motions are completely erratic at equilibrium, albeit they acquire a privileged direction and are thus more ordered out of equilibrium. This happens because of the time asymmetric selection of initial conditions under nonequilibrium conditions, resulting into an invariant probability distribution which breaks the time reversal symmetry by favoring some trajectories with respect to their time reversal images. The temporal ordering is possible at the expense of the increase of the phase-space disorder and is thus compatible with Boltzmann's interpretation of the second law.

The time asymmetry of temporal disorder has been observed for driven Brownian motion in the aforementioned experiments.[11,12] Figure 5c depicts the forward and reversed randomnesses versus the rescaled resolution $\delta = \varepsilon/\sqrt{1 - \exp(-2\tau/\tau_{\mathrm{R}})}$ where $\tau_{\mathrm{R}} \simeq 3$ ms is the relaxation time of the Brownian particle in the laser trap. First of all, we observe that these quantities increase as the resolution goes down to the scale of nanometers, reached thanks to the interferometric techniques.[11,12] This means that the stochastic process of Brownian motion generates more and more temporal disorder as the process is observed on smaller and smaller scales ε. For small values of the spatial resolution ε, we find that

$$h(\varepsilon, \tau) = \frac{1}{\tau} \ln \sqrt{\frac{\pi \mathrm{e} \mathcal{D} \tau_{\mathrm{R}}}{2\varepsilon^2} \left(1 - \mathrm{e}^{-2\tau/\tau_{\mathrm{R}}}\right)} + O(\varepsilon^2) \tag{21}$$

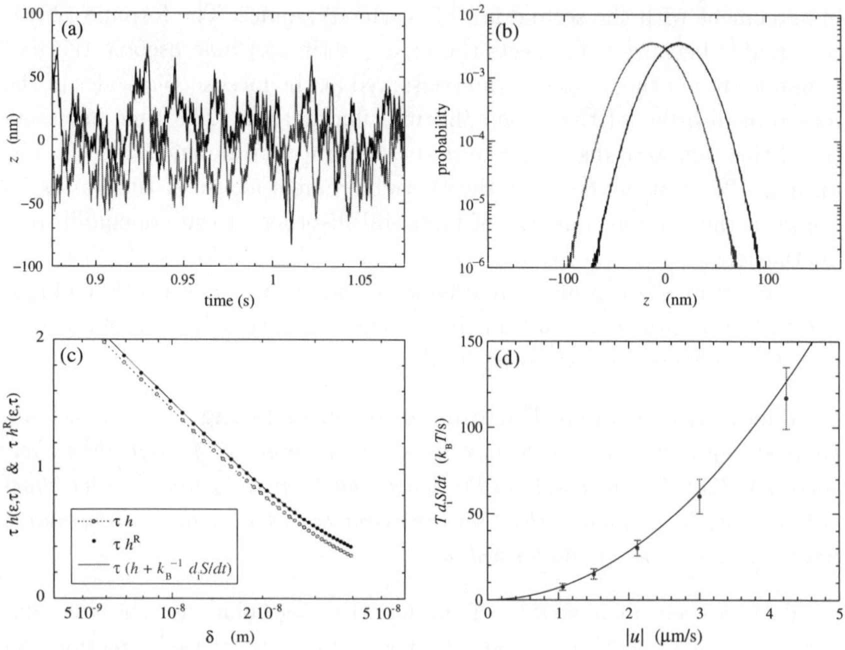

Fig. 5. (a) The time series of a typical path z_t of a trapped Brownian particle in a fluid moving at the speed u for the forward process (upper curve) and $-u$ for the reversed process (lower curve) with $u = 4.24 \times 10^{-6}$ m/s. The temperature is $T = 298$ K and the sampling frequency $f = 8192$ Hz. (b) Gaussian probability distributions of the forward and backward experiments. The mean value is located at $\pm u\tau_{\rm R} = \pm 12.9$ nm. (c) The rescaled forward and reversed randomnesses $\tau h(\varepsilon, \tau)$ and $\tau h^{\rm R}(\varepsilon, \tau)$ for $\tau = 4/f$ versus $\delta = \varepsilon/\sqrt{1 - \exp(-2\tau/\tau_{\rm R})}$. The solid line is the result expected from Eq. (20), showing the good agreement. (d) Thermodynamic entropy production of the Brownian particle versus the driving speed $|u|$. The solid line is the well-known rate of dissipation given by $d_{\rm i}S/dt = \alpha u^2/T$ where α is the viscous friction and T the temperature. The dots depict the difference $h^{\rm R} - h$ between the randomnesses. The formula (20) is thus verified up to experimental errors. The equilibrium state is at zero speed $u = 0$ where the entropy production vanishes. Adapted from Ref.[12]

where $\mathcal{D} = 1.4 \times 10^{-13}\,{\rm m}^2/{\rm s}$ is the diffusion coefficient of the Brownian particle.[12] The second important observation in Fig. 5c is that the reversed randomness $h^{\rm R}$ is larger than the forward randomness h by an amount equal to the thermodynamic entropy production in units of Boltzmann's constant. This is the manifestation of the time asymmetry of nonequilibrium fluctuations, which is here remarkably observed down to the nanoscale.[12] The difference between the reversed and forward randomnesses is quadratic in the fluid speed u as expected from viscous dissipation (see Fig. 5d).

Similar results have been obtained for an RC electric circuit, showing the time asymmetry down to fluctuations of about a few thousands electronic charges.[12] The link between the formula (20) and the escape-rate theory has been discussed elsewhere.[4] The connection can also be established with nonequilibrium work relations.[16,17]

5. The Fluctuation Theorem

Another newly discovered relationship is the fluctuation theorem which concerns the probability distribution of fluctuating quantities such as the currents crossing a nonequilibrium system, the corresponding dissipated heat, or the work performed on the system. Several versions of the fluctuation theorem have been derived for either dynamical systems or stochastic processes.[18–23] The fluctuation theorem is based on the microreversibility implying a symmetry relation between the probabilities of opposite fluctuations.

Recently, we have carried out a derivation of the fluctuation theorem for stochastic processes in the framework of the graph theory by Hill, Schnakenberg, and others.[24–26] This approach allows us to identify the thermodynamic forces – i.e., the De Donder affinities[27,28] – driving the system out of equilibrium.[29–31] In chemical kinetics, the affinities are given in terms of the free enthalpy changes of the reactions:

$$A_\gamma = \frac{\Delta G_\gamma}{k_{\rm B}T} = \frac{G_\gamma - G_{\gamma,{\rm eq}}}{k_{\rm B}T} \tag{22}$$

where T is the temperature.[27,28] In electric devices, the affinities are given by the difference of electronic chemical potentials between the electromotive sources of the circuit. If the affinities vanish, the system returns to equilibrium. Therefore, these affinities drive nonequilibrium currents, which take fluctuating instantaneous values $j_\gamma(t)$. The average of such a current over a finite time interval t is thus the random variable:

$$J_\gamma = \frac{1}{t}\int_0^t j_\gamma(t')\,dt' \tag{23}$$

In chemical reactions, these currents are the rates of the reactive events, i.e., the rates of the transformations of reactants into products. In chemomechanical systems such as the F_1-ATPase rotary molecular motor, a current may represent the number of revolutions per unit time while another is the rate of consumption or synthesis of ATP. Accordingly, the system may sustain several currents $\mathbf{J} = (J_1, J_2, ..., J_\gamma, ..., J_c)$ driven by as many corresponding affinities $\mathbf{A} = (A_1, A_2, ..., A_\gamma, ..., A_c)$. For the molecular motor, these affinities are respectively the external torque acting on its shaft and the affinity of ATP hydrolysis (see below).

The fluctuation theorem for the currents asserts that the ratio of the probabilities of opposite fluctuations goes exponentially with the time interval and the magnitude of the affinities and fluctuations:

$$\frac{P(\mathbf{J})}{P(-\mathbf{J})} \simeq \exp\left(\mathbf{A} \cdot \mathbf{J}\, t\right) \qquad \text{for} \quad t \to \infty \tag{24}$$

The statistical average of the argument in the exponential is nothing else as the thermodynamic entropy production:

$$\frac{1}{k_{\mathrm{B}}} \frac{d_{\mathrm{i}}S}{dt} = \mathbf{A} \cdot \langle \mathbf{J} \rangle \geq 0 \tag{25}$$

The fluctuation theorem for the currents allow us to obtain the generalizations of Onsager's reciprocity relations to the nonlinear response coefficients.[29,31]

If we introduce the decay rate of the probability as $P(\mathbf{J}) \sim \exp\left[-H(\mathbf{J})\, t\right]$, the fluctuation theorem can be expressed in the form:

$$\mathbf{A} \cdot \mathbf{J} = H(-\mathbf{J}) - H(\mathbf{J}) \tag{26}$$

which shows the similarity with the other new relationships that are the equations (13) and (20), as well as the escape-rate formulae.[9,10] If Eq. (26) is evaluated for the mean currents, the thermodynamic entropy production turns out to be given by $H(-\langle \mathbf{J} \rangle)$ since $H(\langle \mathbf{J} \rangle) = 0$. Comparing with Eq. (20), we infer that the reversed randomness is larger – and typically much larger – than the decay rate of the opposite current probability: $h^{\mathrm{R}}(\varepsilon, \tau) \geq H(-\langle \mathbf{J} \rangle)$. This shows that the probability distribution of the currents characterizes the system in a coarser way than the (ε, τ)-entropies per unit time, which can even probe the fluctuations down to the nanoscale.[12] In this sense, the fluctuation theorem is closer to the macroscale than the relationship (20) described in the previous section.

6. Molecular Motors

The new advances reported in the previous sections apply to out-of-equilibrium nanosystems such as the molecular motors, characterizing the way they function in the presence of fluctuations. One of the best known molecular motors is the F_1-ATPase studied by Professor Kinosita and coworkers.[32,33] The F_1-ATPase protein complex is a barrel composed of three large α- and β-subunits circularly arranged around a smaller γ-subunit playing the role of the shaft. The three β-subunits contain the reactive sites for the hydrolysis of ATP. A bead of 40 nm-diameter is attached to the shaft to observe its motion, which clearly shows that the rotation takes place in six substeps: ATP binding induces a rotation of about 90° followed by the release of ADP and P_i during a rotation of about 30° (see Fig. 6). Therefore, the hydrolysis of one ATP corresponds to a rotation by 120° and a revolution of 360° to three sequential ATP hydrolysis in the three β-subunits. There is thus a tight coupling between rotation and chemical reactions in this chemomechanical system.

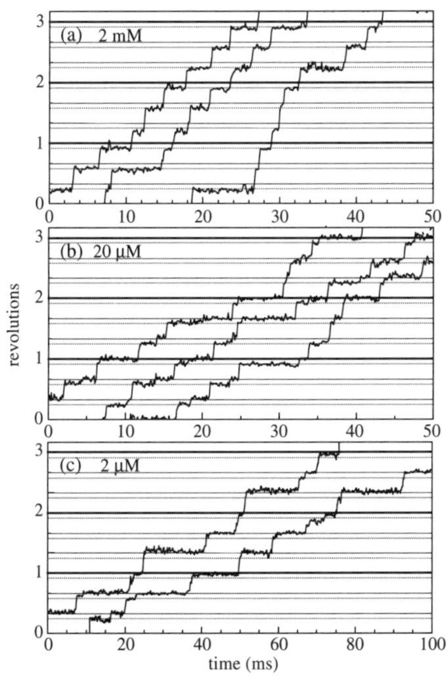

Fig. 6. Experimental data of R. Yasuda, H. Noji, M. Yoshida, K. Kinosita Jr., and H. Itoh[32] on the time courses of stepping rotation of the F_1 motor with a 40-nm bead attached to the rotating γ-subunit at: (a) [ATP] = 2 mM; (b) [ATP] = 20 μM; (c) [ATP] = 2 μM.

In the simplest approximation, the motor can be modeled as a stochastic process with six conformational states whether the sites of the three β-subunits are occupied or not.[34] The random transitions between these states

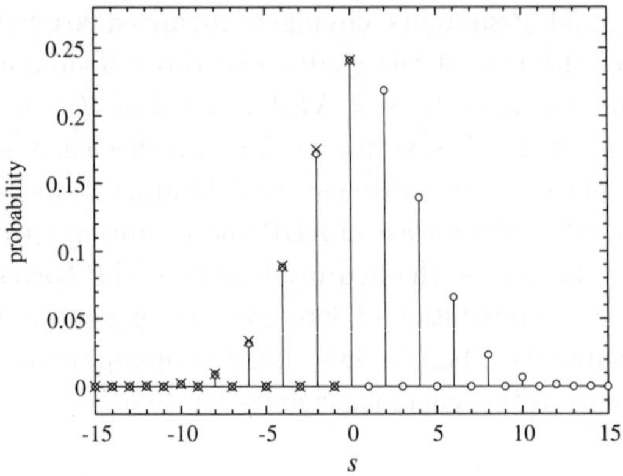

Fig. 7. Probability $P(s)$ (open circles) that the F_1 motor performs s substeps during the time interval $t = 10^4$ s compared with the prediction $P(-s)\exp(sA/6)$ (crosses) of the fluctuation theorem (29) for [ATP] $= 6\ 10^{-8}$ M and [ADP][P$_i$] $= 10^{-2}$ M^2 corresponding to the affinity $A \simeq 0.6$ close to equilibrium.[34]

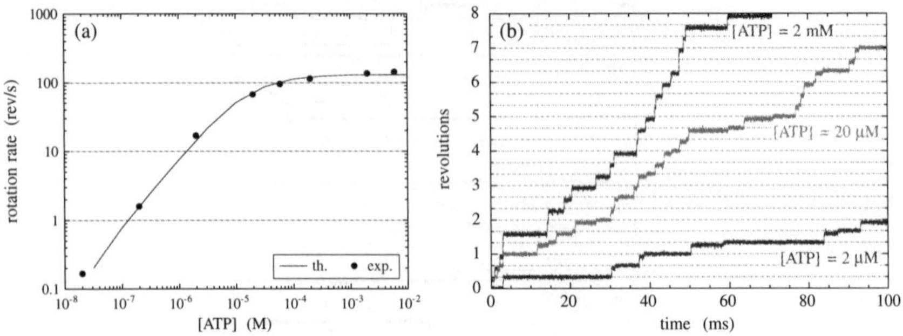

Fig. 8. (a) Rotation rate of the F_1 motor in revolutions per second versus the ATP concentration, under the experimental conditions [ADP][P$_i$] $= 0$, 23°C, and a bead of 40-nm diameter. The dots are the experimental data of Ref.[32] while the solid line is the result of the model of Ref.[35] (b) Stochastic trajectories of the rotation of the F_1 motor simulated by the model of Ref.[35] under the same conditions as the experimental observations of Fig. 6.

are due to the reactive events of the following chemical scheme:

$$\mathrm{ATP} + \mathrm{F}_1(\sigma) \rightleftharpoons \mathrm{F}_1(\sigma+1) \rightleftharpoons \mathrm{F}_1(\sigma+2) + \mathrm{ADP} + \mathrm{P_i} \quad (\sigma = 1,3,5) \quad (27)$$

with $\mathrm{F}_1(7) \equiv \mathrm{F}_1(1)$. The affinity driving the motor out of equilibrium is controlled by the concentrations of ATP and the products of hydrolysis according to

$$A = -3\,\frac{\Delta G^0}{k_{\rm B}T} + 3\,\ln\frac{[\text{ATP}]}{[\text{ADP}][\text{P}_\text{i}]} \tag{28}$$

in terms of the standard free enthalpy of hydrolysis $\Delta G^0 \simeq -50$ pN nm at pH 7 and 23°C.[33] The affinity vanishes at equilibrium where the concentrations reach their equilibrium ratio. Out of equilibrium, the affinity drives the rotation of the motor. The chemical scheme (27) shows that the mean rotation rate has a Michaelis-Menten dependence on the ATP concentration. Accordingly, the F_1 motor typically functions in a highly nonlinear regime far from equilibrium.[34]

Thanks to the fluctuation theorem (24), we can determine the probability of backward rotation as the motor is driven away from equilibrium. We find that[34]

$$\frac{P(s)}{P(-s)} \simeq \exp\left(\frac{sA}{6}\right) \tag{29}$$

for the probability $P(s)$ of a rotation by a signed number s of substeps over a time interval t. A full forward or backward revolution happens when $s = \pm 6$. This fluctuation theorem shows that the probability of backward rotation is reduced by the exponential factor $\exp(-sA/6)$ with respect to the probability of forward rotation. Figure 7 compares the forward and backward probabilities in a situation relatively close to chemical equilibrium, showing the agreement with the prediction of the fluctuation theorem. However, because of the large value of the standard free enthalpy of hydrolysis, the backward substeps rapidly become very rare as the motor is driven in the highly nonlinear regime with $40 < A < 60$ corresponding to the physiological conditions. This result shows that functioning away from equilibrium provides robustness to the biomolecular processes, allowing unidirectional motions to overwhelm erratic motions. The basic mechanism is that the nonequilibrium constraints suppress the time-reversed paths responsible for the erratic motions leaving time asymmetric unidirectional motions, as explained in the previous sections.

The aforementioned model of the F_1 motor can be extended to include the continuous variation of the rotation angle, still keeping the six states of the chemical scheme (27).[35] In this way, it is possible to reproduce not only the Michaelis-Menten dependence of the rotation rate on the ATP concentration, but also the substeps of the stochastic trajectories with great realism (see Fig. 8).

7. Conclusion and Perspectives

Recent progress in nonequilibrium statistical mechanics has achieved the integration of thermodynamics with stochastic aspects in systems driven out of equilibrium. In this regard, the new results provide the basis for a nonequilibrium statistical thermodynamics of nanosystems bridging the gap between the emergent macroscopic phenomena and the molecular motions at the nanoscale. The integration of thermodynamic and stochastic aspects has been made possible thanks to the discovery of new relationships which all share the same mathematical scheme in which an irreversible thermodynamic property is given by the difference between two quantities characterizing the randomness of molecular motions. This scheme appears in the escape-rate formulae,[9,10] Eqs. (13) and (20), as well as the form (26) of the fluctuation theorem. In this way, the thermodynamic entropy production can nowadays be understood as a time asymmetry in the temporal disorder of nonequilibrium fluctuations.

These relationships concern the unidirectional motions of nonequilibrium systems such as the driven Brownian motion and the F_1-ATPase molecular motor. Although their motions are erratic at equilibrium where opposite random steps have equal probabilities, they become unidirectional out of equilibrium by the time asymmetric suppression of backward steps, as explained thanks to the new relationships. These results open important perspectives in our understanding of biological phenomena on the basis of thermodynamics. Indeed, one of the main features of life is metabolism, which is the evidence of the many nonequilibrium processes taking place in cells. The recent results explain that this nonequilibrium regime is responsible for temporal ordering in the behavior of biosystems. In this way, modern statistical thermodynamics can give an answer to the question of the origins of order and information in biological systems.

Acknowledgments

This research is financially supported by the F.R.S.-FNRS Belgium and the "Communauté française de Belgique" (contract "Actions de Recherche Concertées" No. 04/09-312).

References

1. I. Prigogine, *Introduction to Thermodynamics of Irreversible Processes* (Wiley, New York, 1967).
2. P. Gaspard, *Chaos, Scattering and Statistical Mechanics* (Cambridge University Press, Cambridge UK, 1998).

3. P. Gaspard, *Physica A* **369**, 201 (2006).
4. P. Gaspard, *Adv. Chem. Phys.* **135**, 83 (2007).
5. P. Gaspard, *J. Stat. Phys.* **68**, 673 (1992).
6. S. Tasaki and P. Gaspard, *J. Stat. Phys.* **81**, 935 (1995).
7. S. Tasaki and P. Gaspard, *Bussei Kenkyu Cond. Matt.* **66**, 23 (1996).
8. P. Gaspard, I. Claus, T. Gilbert, and J. R. Dorfman, *Phys. Rev. Lett.* **86**, 1506 (2001).
9. P. Gaspard and G. Nicolis, *Phys. Rev. Lett.* **65**, 1693 (1990) .
10. J. R. Dorfman and P. Gaspard, *Phys. Rev. E* **51**, 28 (1995).
11. D. Andrieux, P. Gaspard, S. Ciliberto, N. Garnier, S. Joubaud, and A. Petrosyan, *Phys. Rev. Lett.* **98**, 150601 (2007).
12. D. Andrieux, P. Gaspard, S. Ciliberto, N. Garnier, S. Joubaud, and A. Petrosyan, *J. Stat. Mech.: Th. Exp.* P01002 (2008).
13. P. Gaspard and X.-J. Wang, *Phys. Rep.* **235**, 291 (1993).
14. P. Gaspard, *J. Stat. Phys.* **117**, 599 (2004); *Erratum* **126**, 1109 (2006).
15. P. Gaspard, *C. R. Physique* **8**, 598 (2007).
16. C. Jarzynski, *Phys. Rev. E* **73**, 046105 (2006).
17. R. Kawai, J. M. R. Parrondo, and C. Van den Broeck, *Phys. Rev. Lett.* **98**, 080602 (2007).
18. D. J. Evans, E. G. D. Cohen, and G. P. Morriss, *Phys. Rev. Lett.* **71**, 2401 (1993).
19. G. Gallavotti and E. G. D. Cohen, *Phys. Rev. Lett.* **74**, 2694 (1995).
20. J. Kurchan, *J. Phys. A: Math. Gen.* **31**, 3719 (1998).
21. J. L. Lebowitz and H. Spohn, *J. Stat. Phys.* **95**, 333 (1999).
22. C. Maes, *J. Stat. Phys.* **95**, 367 (1999).
23. G. E. Crooks, *Phys. Rev. E* **60**, 2721 (1999).
24. T. L. Hill, *Free Energy Transduction and Biochemical Cycle Kinetics* (Dover, New York, 2005).
25. J. Schnakenberg, *Rev. Mod. Phys.* **48**, 571 (1976).
26. D.-Q. Jiang, M. Qian, and M.-P. Qian, *Mathematical Theory of Nonequilibrium Steady States* (Springer, Berlin, 2004).
27. T. De Donder and P. Van Rysselberghe, *Affinity* (Stanford University Press, Menlo Park CA, 1936).
28. G. Nicolis and I. Prigogine, *Self-Organization in Nonequilibrium Systems* (Wiley, New York, 1977).
29. D. Andrieux and P. Gaspard, *J. Chem. Phys.* **121**, 6167 (2004); *Erratum* **125**, 219902 (2006).
30. D. Andrieux and P. Gaspard, *J. Stat. Phys.* **127**, 107 (2007).
31. D. Andrieux and P. Gaspard, *J. Stat. Mech.: Th. Exp.* P02006 (2007).
32. R. Yasuda, H. Noji, M. Yoshida, K. Kinosita Jr., and H. Itoh, *Nature* **410**, 898 (2001).
33. K. Kinosita Jr., K. Adachi, and H. Itoh, *Annu. Rev. Biophys. Biomol. Struct.* **33**, 245 (2004).
34. D. Andrieux and P. Gaspard, *Phys. Rev. E* **74**, 011906 (2006).
35. P. Gaspard and E. Gerritsma, *J. Theor. Biol.* **247**, 672 (2007).

A MEASUREMENT-BASED PURIFICATION SCHEME AND DECOHERENCE

H. NAKAZATO

Department of Physics, Waseda University,
Tokyo 169-8555, Japan
E-mail: hiromici@waseda.jp

A purification scheme, proposed several years ago and aimed at extracting a pure state from an arbitrary initial mixed state by repeated measurements, is reviewed and the effects of environment on this scheme are clarified.

Keywords: Purification scheme; quantum state manipulation; decoherence.

1. Introduction

In the field of quantum information and technology,[1] covering such subjects as quantum teleportation, quantum computation, quantum cryptography etc., the quantum coherence is believed to play an important and crucial role, because its notion is one of the most striking features of quantum theory and also because the existence or availability of quantum states with full coherence, that is, pure states that allow superpositions among different states, like

$$|\psi\rangle = \alpha|\uparrow\rangle + \beta|\downarrow\rangle, \qquad |\alpha|^2 + |\beta|^2 = 1, \tag{1}$$

in particular, the so-called entangled states in multipartite systems, is the very premise for quantum manipulations. In this field, a quantum bit, that is, a quantum two-level system, called a *qubit*, is the unit of information and can be expressed as a wave function by means of the up and down terminology of spin 1/2 system, like (1). For a two-qubit system, there are four independent states and the following states with particular entanglement, that form a complete orthonormal set, are often used and called the Bell states

$$|\Psi^\pm\rangle = \frac{1}{\sqrt{2}}\left(|\uparrow\downarrow\rangle \pm |\downarrow\uparrow\rangle\right), \quad |\Phi^\pm\rangle = \frac{1}{\sqrt{2}}\left(|\uparrow\uparrow\rangle \pm |\downarrow\downarrow\rangle\right). \tag{2}$$

Many ideas based on the principles of quantum theory have so far been proposed, assuming that the desired quantum states with full quantum coherence, that is, pure states, especially quantum entangled states, can be prepared and manipulated at will. The states with full coherence are considered to be useful resources in quantum technology, but unfortunately, it is known that they are usually rather fragile against external disturbances and easily lose coherence, that is, quantum character is lost and the system becomes classical.

The loss of coherence (decoherence) is best seen in the density matrix as the disappearance of its off-diagonal terms, with possible changes in diagonal terms due to *dissipation*, representing the approach to the equilibrium

$$\begin{aligned}|\psi\rangle\langle\psi| &= (\alpha|\uparrow\rangle + \beta|\downarrow\rangle)\,(\langle\uparrow|\alpha^* + \langle\downarrow|\beta^*)\\ &= |\alpha|^2|\uparrow\rangle\langle\uparrow| + |\beta|^2|\downarrow\rangle\langle\downarrow| + \alpha\beta^*|\uparrow\rangle\langle\downarrow| + \alpha^*\beta|\downarrow\rangle\langle\uparrow|\\ &\longrightarrow |\alpha'|^2|\uparrow\rangle\langle\uparrow| + |\beta'|^2|\downarrow\rangle\langle\downarrow|, \qquad (|\alpha'|^2 + |\beta'|^2 = 1). \end{aligned} \tag{3}$$

The fragility of coherence means that it is not always easy or rather difficult to obtain quantum states with full coherence, pure states or entangled states in particular. This can be understood from the fact that the familiar thermal state is a typical example of decohered or mixed states. We are facing a serious problem of how to prepare pure, especially entangled states, to keep them and, if they are already decohered, to restore the coherence.

This is an important and crucial problem and there are several proposals for the extraction or distillation of pure and entangled states.[1,2] Here, there are two important key notions; one is the so-called *fidelity* that characterizes how close the extracted state is to the desired pure state. The other one is the *yield*, that is, another measure characterizing the probability of obtaining the desired states. Recall that the purification schemes so far proposed are essentially based on an idea of achieving a higher fidelity, but only for a smaller number of ensembles for each step in the purification procedure, in particular, the highest fidelity (unity) is accomplished only in the limit of infinite number of steps but then the yield becomes vanishingly small. It is generally not a simple or an easy matter to accomplish both requirements of high fidelity and high yield simultaneously.

Several years ago, we proposed yet another scheme of purification,[3] which utilizes the effect of repeated measurements and can meet the above requirement of making the highest fidelity and non-vanishing yield compatible. Furthermore, since the basic idea is very simple, we can consider various variants in order to apply it to different physical situations.

Notice, however, that the scheme has so far been considered in the ideal situations where no external disturbances are present. Because we would not be allowed to neglect or at least have to know such effects when we try to implement the scheme to real experiments, the ability of the scheme has to be examined under the influence of decoherence, caused by surrounding "environment." It is assumed here that the environmental effects can be described by a master equation, instead of the unitary evolution for the ideal case. The framework of the scheme is generalized accordingly, just on the basis of this master equation. The central issue to be addressed is to clarify if the scheme still works under the decoherence and to evaluate its ability. We investigate if it is robust enough or fragile against such disturbances.

In the next section, the framework of our purification scheme is outlined in the ideal case, that is in the absence of environmental disturbances. Two very simple examples are presented for illustration. Next, more realistic or non-ideal cases incorporating the effect of decoherence caused by the environment surrounding the system are examined in Sec. 3. The framework is generalized accordingly and two explicit cases are examined as examples for dephasing and dissipative environments in Subsecs. 3.1 and 3.2, respectively. Section 4 is devoted to a summary.

2. Repeated-measurement-based Purification Scheme: Ideal Case

Here, the framework of the repeated-measurement-based purification[3] is outlined in the ideal case. The basic idea is based on the projection postulate in quantum measurement,[4] according to which, after the measurement, the quantum state is to be projected to the very state just measured. The action of measurement is then conveniently represented by a projection operator. Of course, in a rigorous sense, the measurement itself has to be treated as a physical process, i.e., a generalized spectral-decomposition process, where the state after the measurement is to be endowed with a new degree of freedom to represent the result of measurement.[5] In the following, however, for simplicity, projection operators will be used to represent measurements. This is the ordinary story about the quantum measurement, but it should be noted that the effect of measurement can be more profound and far reaching than one may imagine. This observation constitutes the basis of our purification scheme.

Consider that a part of the total system, say X, is measured and confirmed that it is in a particular state. Of course, just after the measurement, we are sure that X is in that particular state, but what happens to the other part of the total system, say A, that is coupled to X by some interaction

Hamiltonian, but is not measured directly? It is shown later that the effect of measurement can transcend to such a part and if the measurement is repeated many times, system A is to be driven to a pure state, irrespective of its initial state that is in general mixed.

Assume that the total Hamiltonian is composed of those for systems X and A and their mutual interaction

$$H = H_{\rm X} + H_{\rm A} + H_{\rm XA}. \tag{4}$$

The initial state can be taken in the factorized form

$$\rho(0) = |\phi\rangle_{\rm X}\langle\phi| \otimes \rho_{\rm A}(0), \tag{5}$$

without loss of generality, for otherwise, we just measure X to prepare it in the state $|\phi\rangle_{\rm X}$. Notice that the initial state of system A, $\rho_{\rm A}(0)$, is left arbitrary and is generally mixed. Then we measure system X after the total system has evolved for τ under the above Hamiltonian (4) and confirm that X is again in the state $|\phi\rangle_{\rm X}$. This process is repeated many times. The question now is what is the state of system A after N repetitions of such measurements.

In order to answer this question, we recall that the dynamics of system A is governed by the unitary evolution for the period τ, $e^{-iH\tau}$, and the measurement represented by the projection operator $\mathcal{O} \equiv |\phi\rangle_{\rm X}\langle\phi| \otimes 1_{\rm A}$ and the state of A is given by a reduced density matrix after tracing out the X degrees of freedom. It is not difficult to see that the reduced density matrix of A after N such measurements is given by

$$\rho_{\rm A}^{(\tau)}(N) = \frac{\big(V_\phi(\tau)\big)^N \rho_{\rm A}(0) \big(V_\phi^\dagger(\tau)\big)^N}{P^{(\tau)}(N)}. \tag{6}$$

Here, V_ϕ is a projected operator of the unitary evolution

$$V_\phi(\tau) \equiv {}_{\rm X}\langle\phi| e^{-iH\tau} |\phi\rangle_{\rm X}. \tag{7}$$

This is an operator acting on A and is not unitary or hermitian either. The denominator is just the normalization factor given by

$$\begin{aligned} P^{(\tau)}(N) &= {\rm Tr}\Big[(\mathcal{O}e^{-iH\tau}\mathcal{O})^N \rho(0) (\mathcal{O}e^{iH\tau}\mathcal{O})^N\Big] \\ &= {\rm Tr}_{\rm A}\Big[\big(V_\phi(\tau)\big)^N \rho_{\rm A}(0) \big(V_\phi^\dagger(\tau)\big)^N\Big] \end{aligned} \tag{8}$$

and expresses the success probability that the system X has been found in state $|\phi\rangle_{\rm X}$ consecutively N times. The division by this factor reflects the fact that we only keep the right outcomes and those events with wrong results other than state $|\phi\rangle_{\rm X}$ are just discarded.

It is now evident that the fate of system A is governed by the operator V_ϕ and so let us introduce its right- and left-eigenvectors, $|u_n\rangle_{\mathrm{A}}$ and ${}_{\mathrm{A}}\langle v_n|$,

$$V_\phi(\tau)|u_n\rangle_{\mathrm{A}} = \lambda_n|u_n\rangle_{\mathrm{A}}, \quad {}_{\mathrm{A}}\langle v_n|V_\phi(\tau) = {}_{\mathrm{A}}\langle v_n|\lambda_n. \tag{9}$$

Since the operator V_ϕ is not hermitian, $V_\phi(\tau) \neq V_\phi^\dagger(\tau)$, the right- and left-eigenstates are not necessarily the same but form a complete orthonormal set in the following sense

$$\sum_n |u_n\rangle_{\mathrm{A}}\langle v_n| = 1_{\mathrm{A}}, \quad {}_{\mathrm{A}}\langle v_n|u_m\rangle_{\mathrm{A}} = \delta_{nm}, \quad {}_{\mathrm{A}}\langle u_n|u_n\rangle_{\mathrm{A}} = 1 \tag{10}$$

and their eigenvalues are complex-valued. Furthermore, since it is a projected operator of a unitary evolution operator (7), the absolute values of the eigenvalues are upper-bounded

$$1 \geq |\lambda_0| \geq |\lambda_1| \geq \cdots \geq 0. \tag{11}$$

Therefore, if the largest in magnitude eigenvalue λ_0 is *discrete* and the corresponding eigenstate is *unique*, the operator V_ϕ to the power N shall be dominated by a single term

$$\left(V_\phi(\tau)\right)^N = \Big(\sum_n \lambda_n|u_n\rangle_{\mathrm{A}}\langle v_n|\Big)^N = \sum_n \lambda_n^N|u_n\rangle_{\mathrm{A}}\langle v_n| \stackrel{\text{large}N}{\longrightarrow} \lambda_0^N|u_0\rangle_{\mathrm{A}}\langle v_0| \tag{12}$$

and the reduced density matrix $\rho_{\mathrm{A}}^{(\tau)}(N)$ will be dominated by a single term, in other words, it will approach a pure state

$$\rho_{\mathrm{A}}^{(\tau)}(N) \stackrel{\text{large }N}{\longrightarrow} |u_0\rangle_{\mathrm{A}}\langle u_0|, \tag{13}$$

with probability

$$P^{(\tau)}(N) \stackrel{\text{large }N}{\longrightarrow} |\lambda_0|^{2N}{}_{\mathrm{A}}\langle v_0|\rho_{\mathrm{A}}(0)|v_0\rangle_{\mathrm{A}}. \tag{14}$$

Thus, we have shown that our procedure can extract a pure state from an arbitrary mixed state, under the above discreteness and uniqueness conditions.

Furthermore, we note that there is a possibility of an optimal purification, that is, a quick or a fewer step purification without losing any probability if we are able to make the following conditions

$$|\lambda_0| = 1 \quad \text{and} \quad |\lambda_n/\lambda_0| \ll 1, \quad \forall n \neq 0 \tag{15}$$

satisfied by adjusting parameters, like the measurement interval τ and those characterizing state $|\phi\rangle_{\mathrm{X}}$. Under the optimal purification, we are able to extract whole pure-state component contained in the initial mixed state $\rho_{\mathrm{A}}(0)$.

2.1. *Example 1: Qubit purification*

In the following, two very simple cases are considered as examples to see how the above purification scheme works. The first case would be the simplest, where two qubits X and A interact through the spin–spin interaction Hamiltonian

$$H = \frac{\omega}{2}(\sigma_z^{\mathrm{X}} + \sigma_z^{\mathrm{A}}) + g(\sigma_+^{\mathrm{X}}\sigma_-^{\mathrm{A}} + \mathrm{h.c.}) \tag{16}$$

and we measure X in the up state $|\uparrow\rangle_{\mathrm{X}}$ every time-interval τ. Here $\sigma_z^{\mathrm{X(A)}}$ and $\sigma_\pm^{\mathrm{X(A)}}$ are the usual Pauli matrices for qubit X(A) and qubits X and A are treated symmetrically, just for simplicity. If $|\phi\rangle_{\mathrm{X}}$ is set to be $|\uparrow\rangle_{\mathrm{X}}$, the projected operator $V_\phi(\tau) = V_\uparrow(\tau)$ becomes a 2×2 diagonal matrix in the up-down basis and therefore the eigenstates and eigenvalues are easily read from it

$$|u_0\rangle_{\mathrm{A}} = |\uparrow\rangle_{\mathrm{A}}, \quad \lambda_0 = e^{-ig\tau}, \quad |\lambda_0| = 1, \tag{17}$$

$$|u_1\rangle_{\mathrm{A}} = |\downarrow\rangle_{\mathrm{A}}, \quad \lambda_1 = \cos g\tau, \quad |\lambda_1| \le 1. \tag{18}$$

The eigenstate belonging to the largest (in magnitude) eigenvalue is the up state $|\uparrow\rangle_{\mathrm{A}}$ and the magnitude of its eigenvalue is unity. If we make the magnitude of the other eigenvalue less than unity or more efficiently adjust τ to satisfy

$$g\tau = \frac{\pi}{2}, \tag{19}$$

the other eigenvalue λ_1 vanishes identically and we are able to achieve an optimal purification in which every time we measure X to be in the up state, the state of A is projected to the pure up state. In such a case, we need only one measurement of X to extract the pure state in A. This is an extreme case but we can discuss more general cases under the same line of thought.

2.2. *Example 2: Entanglement extraction*

The second example here considered will show us how the scheme allows one to extract an entangled state between noninteracting qubits A and B, by measuring another qubit X that interacts A and B separately. Since the entanglement would be destroyed if the measurement is performed directly on the system (due to the projection postulate), while in this scheme we do not touch the system itself in order to extract an entanglement, the present scheme may be considered to be very much suited for entanglement extraction.

Our model Hamiltonian

$$H = \frac{\omega}{2}(\sigma_z^{\mathrm{A}} + \sigma_z^{\mathrm{B}} + \sigma_z^{\mathrm{X}}) + g(\sigma_+^{\mathrm{A}}\sigma_-^{\mathrm{X}} + \sigma_+^{\mathrm{B}}\sigma_-^{\mathrm{X}} + \text{h.c.}), \tag{20}$$

which is again symmetric between A and B, incorporates the same type of couplings between A and X and B and X as before, while there is no direct interaction between A and B. In order to estimate the projected operator $V_\phi(\tau)$, let us first list up all the eigenstates of the Hamiltonian H. Notice that there are two conserved quantities in this system, that is, the parity under the exchange of A and B and the total number of spin-ups in A+B+X, by which we can easily find all eigenstates. There are six A–B symmetric eigenstates

$$\begin{aligned}
E_1 &= \tfrac{3}{2}\omega, & |E_1\rangle &= \frac{1}{\sqrt{2}}\big[\Phi^+\!\uparrow\rangle + |\Phi^-\!\uparrow\rangle\big],\\
E_{2,3} &= -\tfrac{1}{2}\omega \pm \sqrt{2}g, & |E_{2,3}\rangle &= \frac{1}{2}\big[\Phi^+\!\uparrow\rangle - |\Phi^-\!\uparrow\rangle \pm \sqrt{2}|\Psi^+\!\downarrow\rangle\big],\\
E_4 &= -\tfrac{3}{2}\omega, & |E_4\rangle &= \frac{1}{\sqrt{2}}\big[|\Phi^+\!\downarrow\rangle - |\Phi^-\!\downarrow\rangle\big],\\
E_{5,6} &= \tfrac{1}{2}\omega \pm \sqrt{2}g, & |E_{5,6}\rangle &= \frac{1}{2}\big[|\Phi^+\!\downarrow\rangle + |\Phi^-\!\downarrow\rangle \pm \sqrt{2}|\Psi^+\!\uparrow\rangle\big],
\end{aligned}$$

and two anti-symmetric ones

$$\begin{aligned}
E_7 &= \tfrac{1}{2}\omega, & |E_7\rangle &= |\Psi^-\!\uparrow\rangle,\\
E_8 &= -\tfrac{1}{2}\omega, & |E_8\rangle &= |\Psi^-\!\downarrow\rangle,
\end{aligned}$$

where states of system A+B are expressed in terms of the Bell states $|\Psi^\pm\rangle$ and $|\Phi^\pm\rangle$ (see Eq. (2)). Notice that $|\Psi^-\rangle$ is the only anti-symmetric state among four Bell states.

Since the evolution operator is decomposed in terms of these eigenstates, if we measure the state $|\phi\rangle_{\mathrm{X}}$

$$|\phi\rangle_{\mathrm{X}} = \alpha|\!\uparrow\rangle_{\mathrm{X}} + \beta|\!\downarrow\rangle_{\mathrm{X}}, \quad |\alpha|^2 + |\beta|^2 = 1, \tag{21}$$

in X, the projected operator $V_\phi(\tau)$ reads as

$$\begin{aligned}
V_\phi(\tau) &= {}_{\mathrm{X}}\langle\phi|e^{-iH\tau}|\phi\rangle_{\mathrm{X}} = \sum_{k=1}^{8} e^{-iE_k\tau}{}_{\mathrm{X}}\langle\phi|E_k\rangle\langle E_k|\phi\rangle_{\mathrm{X}}\\
&= \big[|\alpha|^2 e^{-i\omega\tau/2} + |\beta|^2 e^{i\omega\tau/2}\big]|\Psi^-\rangle\langle\Psi^-| + \text{symm. states}.
\end{aligned} \tag{22}$$

Observe that the state $|\Psi^-\rangle$ is separated from the rest and is nothing but one of the eigenstates of $V_\phi(\tau)$ for any state $|\phi\rangle_{\mathrm{X}}$ (here characterized by α and β). This is essentially due to the original symmetry of the Hamiltonian H under A–B exchange. We can make the factor in front of $|\Psi^-\rangle\langle\Psi^-|$, that is nothing but the eigenvalue corresponding to the eigenstate $|\Psi^-\rangle$, unity (in magnitude), just by choosing

$$\omega\tau = 2\pi. \tag{23}$$

Then the state $|\Psi^-\rangle$ becomes the eigenstate belonging to the largest eigenvalue λ_0

$$|\lambda_0| = 1, \quad |u_0\rangle_{\mathrm{AB}} = |\Psi^-\rangle \tag{24}$$

and there is a possibility of extracting this maximally entangled state optimally, provided other eigenvalues are made small compared with λ_0. This is in fact possible: for example, if we choose

$$g\tau = \frac{\pi}{2\sqrt{2}}, \quad |\alpha\beta|^2 = \frac{4}{27}, \quad (|\alpha|^2 + |\beta|^2 = 1),$$

we can obtain analytically all eigenvalues

$$\lambda_0 = -1, \quad \lambda_1 = \frac{2}{3}, \quad \lambda_{2,3} = \frac{1 \pm i\sqrt{7}}{6},$$

which show explicitly that the condition

$$1 = |\lambda_0| > |\lambda_i|, \quad \forall i \neq 0$$

is satisfied. In this case, that is, if we can adjust the measurement interval τ, the coupling constant g or energy level ω, and measure X-spin along a specific direction characterized by the above α and β, the Bell state $|\Psi^-\rangle$ is certainly and gradually extracted from the initial mixed state without any loss. An optimal extraction of the entangled state has thus been explicitly demonstrated.

Needless to say, we usually do not need to worry about making such fine tunings as above in order to extract a pure state. This is because it is generally much more difficult to make two (or more) eingenvalues maximum in magnitude at the same time. Instead, we would have to repeat measurements as many times as necessary, if the corresponding eigenvalue has successfully been made unity in magnitude by a proper choice of parameters, like (23).

Finally, it is to be noted that thanks to the simplicity of the basic idea underlying this scheme, many applications are possible with slight changes to the original. Such applications and investigations include (1) an attempt to implement the scheme to a real physical system, which is in progress in collaboration with an experimental group at Hokkaido University,[6] (2) introduction of a notion of "flying mediator"[7] X to establish entanglement between spatially separated systems (note that in the above example of entanglement extraction, both A and B are assumed to simultaneously interact with X and therefore they can not be separated spatially), (3) a possible construction of a kind of quantum filter when degeneracies are remained in the spectrum of the projected operator,[8] (4) a possibility of extraction of pure states through steady-state scattering processes, where there is no such a notion of interaction time existing, which is automatically taken into account by the setup of the initial state and the scattering potential,[9] and so on.

3. Incorporation of Decoherence

It has been assumed so far that the dynamics of the system X+A is given by the Hamiltonian and therefore its evolution is unitary, but in reality no completely isolated systems are existing physically and we have to duly take into account the presence of environmental degrees of freedom. This is important if we try to implement the scheme into experiments at laboratories. The question is whether the scheme can somehow work even under the external disturbances. Can it be robust enough or fragile against such disturbances?

In order to answer these questions, we have to reconsider everything from the beginning because the dynamics is essentially different from the previous one. In this section, a master equation for the system under consideration is first derived from a microscopic Hamiltonian that models the environment as a collection of bosonic harmonic oscillators, in the ordinary Born-Markov limit.[10] Under this dynamics, the construction and generalization of the framework will be straightforward: one has to take into account the action of measurement by now a projection superoperator and derive the reduced density matrix by tracing out the irrelevant degrees of freedom, just like in the ideal case.

We consider that the dynamics of the system X+A is given by a master equation of the form

$$\dot{\rho} = \mathcal{L}\rho, \quad \mathcal{L} = \mathcal{H} + \mathcal{D}, \tag{25}$$

when it suffers from decoherence, or in other words, if the system X+A is not closed but open. ρ is the density matrix of the system X+A and its time evolution is given by a superoperator $\mathcal{L}$, that includes not only the unitary part $\mathcal{H}$

$$\mathcal{H}\rho = -i[H, \rho], \tag{26}$$

but also a dissipative part $\mathcal{D}$, called a dissipater, that embodies the effect of environment. Formally, this equation is solved in terms of a time-evolution superoperator

$$\mathcal{M}(\tau) = e^{\mathcal{L}\tau}. \tag{27}$$

If the action of measurement is represented by a projection superoperator

$$\mathcal{P}\rho = |\phi\rangle_{\mathrm{X}}\langle\phi|\rho|\phi\rangle_{\mathrm{X}}\langle\phi| \tag{28}$$

that maps the density operator to another density operator, but is essentially the same as the projection operator $\mathcal{O}$ in the ideal case, after the N repetitions of the measurement and the evolution for time interval τ, the density matrix of system X+A, $\rho^{(\tau)}(N)$, is expressed as

$$\begin{aligned}\rho^{(\tau)}(N) &\propto \mathcal{P}\mathcal{M}(\tau)\cdots\mathcal{P}\mathcal{M}(\tau)\mathcal{P}\rho(0) = (\mathcal{P}\mathcal{M}(\tau)\mathcal{P})^N\rho(0)\\ &\equiv (\mathcal{V}(\tau))^N\rho(0).\end{aligned} \tag{29}$$

Since the state of system X is projected on $|\phi\rangle_{\mathrm{X}}$ after the combined action of $\mathcal{P}\mathcal{M}(\tau)$, the above superoperator $\mathcal{V}(\tau)$ is decomposed as

$$\mathcal{V}(\tau) = \mathcal{V}_{\mathrm{A}}(\tau)\mathcal{P}, \tag{30}$$

where the action on system A is represented by a superoperator $\mathcal{V}_{\mathrm{A}}(\tau)$. In order to know its structure, we need more information about the master equation.

Fortunately, we know that the solution of the standard master equation, the one that we are considering here and has to be referred to as the Gorini-Kossakowski-Sudarshan–Lindblad (GKS-L) equation,[11] is always written in the form

$$\mathcal{M}(\tau)\rho = \sum_{k=0}^{G} T_k(\tau)\,\rho\,T_k^\dagger(\tau), \quad \sum_{k=0}^{G} T_k^\dagger(\tau)T_k(\tau) = 1, \tag{31}$$

which shall be referred to as the Sudarshan-Mathews-Rau–Kraus (SMR-K) representation.[12] Therefore the superoperator $\mathcal{V}_{\mathrm{A}}(\tau)$ is given by a summation of operations

$$\begin{aligned}\mathcal{V}_{\mathrm{A}}(\tau)\rho_{\mathrm{A}} &= \sum_{k=0}^{G} {}_{\mathrm{X}}\langle\phi|T_k(\tau)|\phi\rangle_{\mathrm{X}}\ \rho_{\mathrm{A}}\ {}_{\mathrm{X}}\langle\phi|T_k^\dagger(\tau)|\phi\rangle_{\mathrm{X}}\\ &\equiv \sum_{k=0}^{G} V_k(\tau)\,\rho_{\mathrm{A}} V_k^\dagger(\tau).\end{aligned} \tag{32}$$

Notice that this type of operation has also appeared in the ideal case but there is no summation over k and only a single term $V_\phi(\tau)$ appears. This is one of the essential differences between ideal and non-ideal cases. In other words, the presence of dissipater $\mathcal{D}$, that is reflecting the presence of external environment, manifests in the fact that there are more than one operators V_k involved in the projected dynamics, while in the ideal case, only a single operator corresponding to the unitary evolution appears.

It is evident that the superoperator $\mathcal{V}_{\mathrm{A}}(\tau)$ decides the fate of system A and so even though not much has been known about superoperators in general, we can proceed formally like in the ideal case. If it is decomposed in terms of its eigenprojections Π_n

$$\mathcal{V}_{\mathrm{A}}(\tau) = \sum_n \lambda_n \Pi_n, \quad \Pi_n \Pi_m = \delta_{nm} \Pi_n, \quad \sum_n \Pi_n = 1 \tag{33}$$

and if the largest (in magnitude) eigenvalue λ_0 is discrete and unique, we can expect that the eigenspace corresponding to the eigenprojection Π_0 will be extracted for large N

$$(\mathcal{V}_{\mathrm{A}}(\tau))^N = \sum_n \lambda_n^N \Pi_n \longrightarrow \lambda_0^N \Pi_0. \tag{34}$$

Unfortunately, these statements remain at a rather formal level and not much can be drawn from them. Instead, we can prove the following more practical statement:[13]

> *For a pure state to be an eigenstate of the superoperator $\mathcal{V}_{\mathrm{A}}(\tau)$, it should be a simultaneous eigenstate of all the operators $V_k(\tau)$ involved in (32) and vice versa.* (35)

Remember that there is only a single term, $V_0(\tau) \equiv V_\phi(\tau)$, in the ideal case. This statement implies that it would be difficult to find an eigenstate that is pure, not mixed, of the superoperator in non-ideal cases with $G > 0$. We thus anticipate that it might be difficult to purify the system by this scheme when it is under the influence of decohering environment. In order to get more insight, consider two specific cases below.

3.1. *Purification under dephasing environment*

Let the total system be composed of two qubits X and A, like in the ideal case, but they are assumed to suffer from an environmental influence. The Hamiltonian H for X+A is the same as before but both X and A are interacting with bosonic oscillators modeling the environment. The total

Hamiltonian reads as

$$H_{\rm T} = H + H_{\rm b} + H_{\rm int}, \tag{36}$$

$$H = \frac{\omega}{2}(\sigma_z^{\rm X} + \sigma_z^{\rm A}) + g(\sigma_+^{\rm X}\sigma_-^{\rm A} + \text{h.c.}), \tag{37}$$

$$H_{\rm b} = \int d\boldsymbol{k}\, \omega_{\boldsymbol{k}} a_{\boldsymbol{k}}^\dagger a_{\boldsymbol{k}}, \tag{38}$$

$$H_{\rm int} = (\sigma_z^{\rm X} + \sigma_z^{\rm A}) \int d\boldsymbol{k}\, \big(h(\boldsymbol{k}) a_{\boldsymbol{k}} + \text{h.c.}\big). \tag{39}$$

Observe that this environmental interaction $H_{\rm int}$ does not cause any transitions between different levels and disturbs only the phases of states of X and A. This is why it is called a *dephasing* interaction. Incidentally, for simplicity, X and A are treated symmetrically.

It is straightforward to derive a GKS-L master equation

$$\dot{\rho} = \mathcal{L}\rho, \quad \mathcal{L} = \mathcal{H} + \mathcal{D}, \quad \mathcal{H}\rho = -i[H, \rho] \tag{40}$$

with a dissipater $\mathcal{D}$, explicitly given by

$$\mathcal{D}\rho = \gamma\big[(|2\rangle\langle 2| - |0\rangle\langle 0|)\rho(|2\rangle\langle 2| - |0\rangle\langle 0|) - \frac{1}{2}\{|2\rangle\langle 2| + |0\rangle\langle 0|, \rho\}\big]. \tag{41}$$

Here γ characterizes the dephasing rate and is related to the mode function $h(\boldsymbol{k})$ as usual (see, e.g., Facchi *et al.*[10]) and the states $|2\rangle$ and $|0\rangle$ are two of the (four) eigenstates of the Hamiltonian H, which read as

$$\begin{aligned} E_2 &= \omega + \Delta, & |2\rangle &= |\uparrow\uparrow\rangle, \\ E_0 &= -\omega + \Delta, & |0\rangle &= |\downarrow\downarrow\rangle, \\ E_\pm &= \pm g, & |\Psi^\pm\rangle &= \frac{1}{\sqrt{2}}[|\uparrow\downarrow\rangle \pm |\downarrow\uparrow\rangle], \end{aligned}$$

where an energy shift due to the environmental interaction Δ has already been incorporated in the energy eigenvalues E_i ($i = 2, 0, \pm$). The above master equation (40) with (41) is solved in the standard way[14] and its solution is given in the SMR-K representation

$$\rho(\tau) = \mathcal{M}(\tau)\rho(0) = \sum_{k=0,1,2} T_k(\tau)\,\rho(0)\,T_k^\dagger(\tau). \tag{42}$$

There are three operators $T_0(\tau)$, $T_1(\tau)$ and $T_2(\tau)$ involved

$$\begin{aligned} T_0(\tau) &= e^{-iE_2 t - \gamma\tau/2}|2\rangle\langle 2| + e^{-iE_0\tau - \gamma\tau/2}|0\rangle\langle 0| \\ &\qquad + e^{-iE_+\tau}|\Psi^+\rangle\langle\Psi^+| + e^{-iE_-\tau}|\Psi^-\rangle\langle\Psi^-|, \\ T_1(\tau) &= \sqrt{\cosh\gamma\tau - 1}\big[e^{-iE_2\tau - \gamma\tau/2}|2\rangle\langle 2| + e^{-iE_0\tau - \gamma\tau/2}|0\rangle\langle 0|\big], \\ T_2(\tau) &= \sqrt{\sinh\gamma\tau}\big[e^{-iE_2\tau - \gamma\tau/2}|2\rangle\langle 2| - e^{-iE_0\tau - \gamma\tau/2}|0\rangle\langle 0|\big]. \end{aligned}$$

Notice that $T_0(\tau)$ reduces to the unitary evolution operator in the $\gamma \to 0$ limit, while the others vanish in the same limit and therefore they are solely due to the presence of dephasing interaction.

According to the spirit of our purification, let us measure the system X and confirm that it is in the up sate, i.e., $|\phi\rangle_{\mathrm{X}} = |\uparrow\rangle_{\mathrm{X}}$, because we have shown in Subsec. 2.1 that repeated confirmations of $|\uparrow\rangle_{\mathrm{X}}$ have resulted in an optimal extraction of the up state $|\uparrow\rangle_{\mathrm{A}}$. The corresponding superoperator $\mathcal{V}_{\mathrm{A}}(\tau)$ is expressed in terms of the three operators $V_0(\tau)$, $V_1(\tau)$ and $V_2(\tau)$,

$$\mathcal{V}_{\mathrm{A}}(\tau)\rho_{\mathrm{A}} = \sum_{k=0,1,2} V_k(\tau)\,\rho_{\mathrm{A}} V_k^\dagger(\tau), \tag{43}$$

that are obtained just by projecting $T_k(\tau)$ $(k = 0,1,2)$ to the up state of X

$$\begin{aligned} V_0(\tau) &= {}_{\mathrm{X}}\langle\uparrow|T_0(\tau)|\uparrow\rangle_{\mathrm{X}} = e^{-iE_2\tau-\gamma\tau/2}|\uparrow\rangle_{\mathrm{A}}\langle\uparrow| + \cos E_+\tau|\downarrow\rangle_{\mathrm{A}}\langle\downarrow|, \\ V_1(\tau) &= {}_{\mathrm{X}}\langle\uparrow|T_1(\tau)|\uparrow\rangle_{\mathrm{X}} = \sqrt{\cosh\gamma\tau - 1}e^{-iE_2\tau-\gamma\tau/2}|\uparrow\rangle_{\mathrm{A}}\langle\uparrow|, \\ V_2(\tau) &= {}_{\mathrm{X}}\langle\uparrow|T_2(\tau)|\uparrow\rangle_{\mathrm{X}} = \sqrt{\sinh\gamma\tau}e^{-iE_2\tau-\gamma\tau/2}|\uparrow\rangle_{\mathrm{A}}\langle\uparrow|. \end{aligned}$$

Interestingly, they are all diagonal in the up-down basis and therefore it is quite easy to understand that their eigenstates are up and down states.

According to the above statement (35), the up and down states are the only pure eigenstates of the superoperator $\mathcal{V}_{\mathrm{A}}(\tau)$ and actually they are, as these relations show

$$\begin{aligned} \mathcal{V}_{\mathrm{A}}(\tau)\big[|\uparrow\rangle_{\mathrm{A}}\langle\uparrow|\big] &= \sum_{k=0,1,2} V_k(\tau)|\uparrow\rangle_{\mathrm{A}}\langle\uparrow|V_k^\dagger(\tau) = |\uparrow\rangle_{\mathrm{A}}\langle\uparrow|, \\ \mathcal{V}_{\mathrm{A}}(\tau)\big[|\downarrow\rangle_{\mathrm{A}}\langle\downarrow|\big] &= \cos^2 E_+\tau\big[|\downarrow\rangle_{\mathrm{A}}\langle\downarrow|\big], \\ \mathcal{V}_{\mathrm{A}}(\tau)\big[|\uparrow\rangle_{\mathrm{A}}\langle\downarrow|\big] &= e^{-iE_2\tau-\gamma\tau/2}\cos E_+\tau\big[|\uparrow\rangle_{\mathrm{A}}\langle\downarrow|\big], \\ \mathcal{V}_{\mathrm{A}}(\tau)\big[|\downarrow\rangle_{\mathrm{A}}\langle\uparrow|\big] &= e^{iE_2\tau-\gamma\tau/2}\cos E_+\tau\big[|\downarrow\rangle_{\mathrm{A}}\langle\uparrow|\big]. \end{aligned}$$

Since the eigenvalue of the up state $|\uparrow\rangle_{\mathrm{A}}\langle\uparrow|$ is unity and that of down $|\downarrow\rangle_{\mathrm{A}}\langle\downarrow|$ is $\cos^2 E_+\tau$, we understand that under just the same condition for τ, e.g, (19), an optimal purification to the up state is still possible. This means that the purification scheme *can work* even under the influence of dephasing environment. The scheme is robust enough against this kind of dephasing environment.

3.2. *Dissipative environment*

Next, consider the same two-qubit system X+A but under another environmental interaction, that is, a dissipative environment that causes transitions between different levels, for example, by the following interaction

Hamiltonian

$$H_{\text{int}} = \int d\boldsymbol{k}\,\big[h(\boldsymbol{k})a_{\boldsymbol{k}}(\sigma_+^{\text{X}} + \sigma_+^{\text{A}}) + \text{h.c.}\big]. \tag{44}$$

It is understood that by lowering (raising) the bosonic state, the levels of X and A are shifted up (down). Under this dissipative environment (at zero temperature, for simplicity), the GKS-L master equation is derived in the standard way and the dissipater $\mathcal{D}$ now reads as

$$\begin{aligned}\mathcal{D}\rho = \gamma_{2\to\Psi^+}&\Big[|\Psi^+\rangle\langle 2|\rho|2\rangle\langle\Psi^+| - \tfrac{1}{2}\{|2\rangle\langle 2|, \rho\}\Big]\\ &+ \gamma_{\Psi^+\to 0}\Big[|0\rangle\langle\Psi^+|\rho|\Psi^+\rangle\langle 0| - \tfrac{1}{2}\{|\Psi^+\rangle\langle\Psi^+|, \rho\}\Big].\end{aligned} \tag{45}$$

There are terms responsible for the transitions $|2\rangle \to |\Psi^+\rangle$ and $|\Psi^+\rangle \to |0\rangle$, represented by the first line and the second line of this equation. These states are again eigenstates of the Hamiltonian H (16) [or (37)]

$$\begin{aligned} E_2 &= \omega + \Delta_2, & |2\rangle &= |\uparrow\uparrow\rangle,\\ E_0 &= -\omega, & |0\rangle &= |\downarrow\downarrow\rangle,\\ E_+ = g + \Delta_+,\ E_- &= -g, & |\Psi^\pm\rangle &= \frac{1}{\sqrt{2}}[|\uparrow\downarrow\rangle \pm |\downarrow\uparrow\rangle],\end{aligned}$$

where the inequality $\omega > g$ has been assumed so that the state $|0\rangle$ is the lowest energy state of the Hamiltonian. Δ_k $(k = 2, +)$ are energy shifts caused by the environmental interaction. Furthermore, in order to simplify the notation, assume that the two decay rates $\gamma_{2\to\Psi^+}$ and $\gamma_{\Psi^+\to 0}$ are the same and are given by a single γ, i.e., $\gamma_{2\to\Psi^+} = \gamma_{\Psi^+\to 0} \equiv \gamma$.

The solution of the GKS-L master equation again falls into the following SMR-K representation

$$\rho(\tau) = \mathcal{M}(\tau)\rho(0) = \sum_{k=0,1,2,3} T_k(\tau)\,\rho(0)\,T_k^\dagger(\tau) \tag{46}$$

and in this case, four operators $T_k(\tau)$ $(k = 0 \sim 4)$ involved are given by

$$\begin{aligned} T_0(\tau) &= e^{-iE_2\tau - \gamma\tau/2}|2\rangle\langle 2| + e^{-iE_0\tau}|0\rangle\langle 0|\\ &\qquad + e^{-iE_+\tau-\gamma\tau/2}|\Psi^+\rangle\langle\Psi^+| + e^{-iE_-\tau}|\Psi^-\rangle\langle\Psi^-|,\\ T_1(\tau) &= \sqrt{1-e^{-\gamma\tau}}\,|0\rangle\langle\Psi^+|,\\ T_2(\tau) &= \sqrt{1-e^{-\gamma\tau}-\gamma\tau e^{-\gamma\tau}}\,|0\rangle\langle 2|,\\ T_3(\tau) &= \sqrt{\gamma\tau e^{-\gamma\tau}}\,|\Psi^+\rangle\langle 2|.\end{aligned}$$

It is evident that the last three operators are responsible for the level transitions, $|\Psi^+\rangle \to |0\rangle$, $|2\rangle \to |0\rangle$ and $|2\rangle \to |\Psi^+\rangle$.

It is not difficult to confirm that if these operators are projected on the up state of system X, the resulting operators $V_k = {}_{\rm X}\langle\uparrow|T_k|\uparrow\rangle_{\rm X}$ do not allow the up state $|\uparrow\rangle_{\rm A}$ to be a simultaneous eigenstate of all V_ks, unlike in the case of dephasing environment. This means that the optimal purification that can be achieved both in the ideal case and in the case under the dephasing environment can no longer be expected in this case of dissipative environment. The latter surely deteriorates the ability of this measurement-based purification scheme.

In order to get more insight, it would be desirable to make the discussion more general and consider that we measure system X to confirm that it is in a state $|\alpha\rangle_{\rm X}$, characterized by a complex number α

$$|\alpha\rangle_{\rm X} \equiv \frac{1}{\sqrt{1+|\alpha|^2}}\Big[\alpha|\uparrow\rangle_{\rm X} + |\downarrow\rangle_{\rm X}\Big]. \tag{47}$$

Obviously $|\uparrow\rangle_{\rm X} = \lim_{\alpha\to\infty}|\alpha\rangle_{\rm X}$. Then the superoperator for system A, $\mathcal{V}_{\rm A}(\tau)$, is shown to be expressed as

$$\mathcal{V}_{\rm A}(\tau)\rho_{\rm A} = V_0(\tau)\rho_{\rm A}V_0^\dagger(\tau) + F(\rho_{\rm A},\tau)|\downarrow\rangle_{\rm A}\langle\downarrow| + G(\rho_{\rm A},\tau)|\alpha^*\rangle_{\rm A}\langle\alpha^*|, \tag{48}$$

where

$$V_0(\tau) = {}_{\rm X}\langle\alpha|T_0(\tau)|\alpha\rangle_{\rm X}, \tag{49}$$

$$|\alpha^*\rangle_{\rm A} = \frac{1}{\sqrt{1+|\alpha|^2}}\Big[|\uparrow\rangle_{\rm A} + \alpha^*|\downarrow\rangle_{\rm A}\Big], \tag{50}$$

and F and G are functions of $\rho_{\rm A}$ and τ. Details[15] of F and G are suppressed here, but it would be evident that after the measurement, system A is driven to a mixture of the down state $|\downarrow\rangle_{\rm A}\langle\downarrow|$, the state $|\alpha^*\rangle_{\rm A}\langle\alpha^*|$ and the eigenstates of $V_0(\tau)$.

The result looks rather complicated but we can say the following thing. The only possible pure eigenstate of this superoperator is the down state $|\downarrow\rangle_{\rm A}\langle\downarrow|$ and no other state can be a pure eigenstate because of the statement (35) shown above. Actually, when the up state $|\uparrow\rangle_{\rm X}$, that is the state $|\alpha\rangle_{\rm X}$ at $\alpha\to\infty$ limit, is measured in X, the state $|\alpha^*\rangle_{\rm A}$ becomes nothing but the down state, $\lim_{\alpha\to\infty}|\alpha^*\rangle_{\rm A} \propto |\downarrow\rangle_{\rm A}$, and we can confirm that the down state is certainly extracted in system A. Unfortunately, the result is not very much interesting, because the state $|\downarrow\rangle_{\rm A}$ is the lowest energy state, even though this is considered to be a successful result of purification of the system A whose equilibrium state is given by a mixed state composed of both the up state $|\uparrow\rangle_{\rm A}\langle\uparrow|$ and the down state $|\downarrow\rangle_{\rm A}\langle\downarrow|$. The result implies that the scheme is not strong enough against dissipation and the dissipation surely deteriorates the ability of the scheme. This is a remarkable contrast to the case of dephasing environment.

Even though we are unable to extract a pure state by just increasing the number of measurements, it seems that there is a kind of competition between the effect of projection, that has a tendency to drive the system to a pure state, and that of dissipation, and therefore at an intermediate number of measurements, we may expect to obtain a pure state as a dominant component in the density matrix of A. Details of such estimation[15] are too complicated to present here, but the moral is we had better not to perform measurements indefinitely, if some imperfections are anticipated.

4. Summary

A new purification scheme has been introduced, that is based on a simple idea that the action of measurement of a part of the system can affect the other part drastically and when the measurement is repeated, a pure state can be extracted in the other part. After formulating the idea in the ideal case, that is, in the absence of external disturbance, and showing the possibility of optimal purifications, the idea is examined in non-ideal situations, that is, under the influence of environment. It is shown that the scheme can be robust against the dephasing effect, while the dissipation surely deteriorates its ability. The reason may be roughly understood in the following way. The scheme heavily depends on the projective action and since the dephasing environment does not cause any state changes, its presence has no effect on the projection and therefore on this scheme, while the dissipative environment certainly brings about state changes and interferes the projective action and it can not be compatible with the scheme. Though these conclusions are based on simple model calculations, if one tries to implement this scheme to real experiments at laboratories, it is advised to take into account possible sources of imperfection and to find optimal conditions.

Acknowledgments

This work has been done in collaboration with some of young colleagues and students at Waseda University and with A. Messina's group in Palermo, Italy. It is partly supported by the bilateral Italian-Japanese project 15C1 on "Quantum Information and Computation" of the Italian Ministry for Foreign Affairs, by a Grant for The 21st Century COE Program (Physics of Self-Organization Systems) at Waseda University from the Ministry of Education, Culture, Sports, Science and Technology, Japan, and by a Grant-in-Aid for Scientific Research (C) (No. 18540292) from the Japan Society for the Promotion of Science.

References

1. For reviews on quantum information, see: *The Physics of Quantum Information*, edited by D. Bouwmeester, A. Zeilinger, and A. Ekert (Springer-Verlag, Berlin, 2000); M. A. Nielsen and I. L. Chuang, *Quantum Computation and Quantum Information* (Cambridge University Press, Cambridge, 2000); C. H. Bennett and D. P. DiVincenzo, *Nature* (London) **404**, 247 (2000); A. Galindo and M. A. Martín-Delgado, *Rev. Mod. Phys.* **74**, 347 (2002).
2. C. H. Bennett, G. Brassard, S. Popescu, B. Schumacher, J. A. Smolin, and W. K. Wootters, *Phys. Rev. Lett.* **76**, 722 (1996);**78**, 2031(E) (1997); C. H. Bennett, D. P. DiVincenzo, J. A. Smolin, and W. K. Wootters, *Phys. Rev. A* **54**, 3824 (1996); J. I. Cirac, A. K. Ekert, and C. Macchiavello, *Phys. Rev. Lett.* **82**, 4344 (1999).
3. H. Nakazato, T. Takazawa, and K. Yuasa, *Phys. Rev. Lett.* **90**, 060401 (2003); H. Nakazato, M. Unoki, and K. Yuasa, *Phys. Rev. A* **70**, 012303 (2004).
4. See, for example, B. d'Espagnat, *Conceptual Foundations of Quantum Mechanics* (Benjamin, New York 1971); *Quantum Theory and Measurement*, edited by J. A. Wheeler and W. H. Zurek (Princeton Univ. Press, Princeton, 1983); P. Busch, P. J. Lahti, and P. Mittelstaedt, *The Quantum Theory of Measurement* (Springer-Verlag, Berlin, 1991); M. Namiki and S. Pascazio, *Phys. Rep.* **232**, 301 (1993).
5. S. Pascazio and M. Namiki, *Phys. Rev. A* **50**, 4582 (1994).
6. K. Yoh, K. Yuasa, and H. Nakazato, *Physica E* **29**, 674 (2005); K. Yuasa, K. Okano, H. Nakazato, S. Kashiwada, and K. Yoh, *AIP Conf. Proc.* **893**, 1109 (2007).
7. G. Compagno, A. Messina, H. Nakazato, A. Napoli, M. Unoki, and K. Yuasa, *Phys. Rev. A* **70**, 052316 (2004); K. Yuasa and H. Nakazato, *Prog. Theor. Phys.* **114**, 523 (2005); *J. Phys. A* **40**, 297 (2007).
8. B. Militello, H. Nakazato, and A. Messina, *Phys. Rev. A*, **71**, 032102 (2005); B. Militello, A. Messina, and H. Nakazato, *Opt. Spectrosc.* **99**, 438 (2005).
9. K. Yuasa and H. Nakazato, *J. Phys. A* **40**, 297 (2007).
10. See, for example: H.-P. Breuer and F. Petruccione, *The Theory of Open Quantum Systems* (Oxford University Press, Oxford, 2002); C. W. Gardiner and P. Zoller, *Quantum Noise*, 3rd edition (Springer, Berlin, 2004); P. Facchi, S. Tasaki, S. Pascazio, H. Nakazato, A. Tokuse, and D. A. Lidar, *Phys. Rev. A* **71**, 022302 (2005).
11. V. Gorini, A. Kossakowski, and E. C. G. Sudarshan, *J. Math. Phys.* **17**, 821 (1976); G. Lindblad, *Commun. Math. Phys.* **48**, 119 (1976).
12. E. C. G. Sudarshan, P. M. Mathews, and J. Rau, *Phys. Rev.* **121**, 920 (1961); K. Kraus, *Ann. Phys.* (N.Y.) **64**, 311 (1971); *States, Effects, and Operations* (Springer-Verlag, Berlin, 1983).

13. B. Militello, K. Yuasa, H. Nakazato, and A. Messina, *Phys. Rev. A* **76**, 042110 (2007).
14. H. Nakazato, Y. Hida, K. Yuasa, B. Militello, A. Napoli, and A. Messina, *Phys. Rev. A* **74**, 062113 (2006).
15. A. Messina, B. Militello, H. Nakazato, and K. Yuasa, `arXiv:0711.2751v1` [quant-ph].

QUANTUM FLUCTUATION THEOREM IN THE EXISTENCE OF THE TUNNELING AND THE THERMAL ACTIVATION

TAKAAKI MONNAI*

Department of Applied Physics, Waseda University,
3-4-1 Okubo, Shinjuku-ku, Tokyo 169-8555, Japan
**E-mail: monnai@suou.waseda.jp*

We explore the fluctuation theorem for a diffusion in a tilted washboard potential where both the quantum tunneling and the thermal activation play a role. The diffusion process is treated as a biased random walk where the transition rates are enhanced due to the quantum tunneling. Within this framework, the dissipation is easy to handle, since we need not to measure the experimentally inaccessible quantity like total Hamiltonian which includes the bath degrees of freedom.

Keywords: Fluctuation theorem; thermal activation; quantum tunneling.

1. Introduction

The second law of thermodynamics states that the average entropy production is positive. In mesoscopic systems, the thermal fluctuation becomes relatively large, thus the entropy production fluctuates and happens to be negative. For a wide class of classical systems, there is a symmetry for the fluctuation of the entropy production called fluctuation theorem.[1–4] Fluctuation theorem contains various other results of nonequilibrium systems such as the second law of thermodynamics,[5,6] the linear and nonlinear response theories,[3,7] and Jarzynski's work theorem.[8–10] On the other hand, the quantum extension of fluctuation theorem still remains as a no-man's land. The entropy production can be estimated by observed values of Hamiltonian operators at initial and final times.[11–14] Alternatively, one can define an operator of work and concern with its spectrum.[15] The difference between these approaches has been well-studied.[16–18] Due to the operator nature of the entropy production, these approaches deal with all the degrees of freedom, i.e., a system plus a reservoir. This fact makes their experimental verification difficult.

In this paper, we propose a quantum fluctuation theorem based on the quantum escape rate theory. We explore the entropy production for a diffusion process in a tilted washboard potential. Remarkably, our result will be experimentally accessible as a phase diffusion of the Josephson junction system.

2. Quantum Escape Rate

In this section, we quickly review the quantum escape rate.[19] The so-called ImF method enables us to estimate the escape rate, because the metastability is attributed to the imaginary part of the eigenenergy.[22] We consider a system linearly coupling to a harmonic reservoir at an inverse temperature β. Such a model leads to a generalized Langevin equation,[20,21] and is suitable to consider the quantum Kramers escape rate. The total Hamiltonian is given as

$$H = \frac{p^2}{2M} + V(q) + \int d\lambda \left(\frac{p_\lambda^2}{2M_\lambda} + \frac{M_\lambda \omega_\lambda^2}{2} (q_\lambda - \frac{c_\lambda}{M_\lambda \omega_\lambda^2} q)^2 \right). \tag{1}$$

Here M and M_λ are the mass of the system coordinate in a tilted L-periodic potential $V(q)$, and that of the λ-th harmonic oscillator of the reservoir, and c_λ is the coupling constant. We shall concern with the escape rate from a metastable well. The quantum partition function $Z = \mathrm{Tr} e^{-\beta H}$ is evaluated by the Euclidean path-integral

$$Z = \int Dq e^{-\frac{1}{\hbar} S_{\text{eff}}}, \tag{2}$$

where the Gaussian integral over the reservoir degrees of freedom leads to the effective action

$$S_{\text{eff}} = \int_0^{\hbar\beta} d\tau \left(\frac{M}{2} \dot{q}^2 + V(q) \right) + \frac{1}{2} \int_0^{\hbar\beta} d\tau \int_0^{\hbar\beta} ds k(\tau - s) q(\tau) q(s). \tag{3}$$

Here the memory kernel $k(\tau)$ is essentially the Laplace transform of the dissipation kernel $\gamma(\tau)$ of the corresponding Langevin equation:

$$k(\tau) = \frac{M}{\hbar\beta} \sum_n |\nu_n| \hat{\gamma}(|\nu_n|) e^{i\nu_n \tau};$$

$$\gamma(\tau) = \int d\lambda \frac{c_\lambda^2}{M M_\lambda \omega_\lambda^2} \cos \omega_\lambda t. \tag{4}$$

Here $\nu_n = \frac{2\pi n}{\hbar\beta}$ is the Matsubara frequency.

The partition function is evaluated by the semiclassical steepest descent approximation. The first-order functional derivative of the action gives the equation of motion of a fictitious particle in an inverted potential $-V(q)$ with a periodic boundary condition $q(0) = q(\beta\hbar)$. The corresponding classical solutions gives dominant contribution to the action. There are stationary solutions at the bottom of a well $q(\tau) = q_m$ and the top of a barrier $q(\tau) = q_M$. At a sufficiently high temperature, the fluctuations around these stationary points dominate the action. The fluctuations around the bottom q_m is taken as the second-order variational of the action, while the fluctuations at the barrier q_M amounts to the imaginary contribution to the partition function. In this way, the escape rates to the neighboring wells in positive and negative directions of perturbation are evaluated as[23–25]

$$\begin{aligned} k_+ &= \frac{\omega_0 \omega_R}{2\pi\omega_b} f_q e^{-\beta(\Delta V - \theta F L)}, \\ k_- &= \frac{\omega_0 \omega_R}{2\pi\omega_b} f_q e^{-\beta(\Delta V + (1-\theta) F L)}, \end{aligned} \tag{5}$$

with the frequency at the well-bottom ω_0, and that at the barrier-top ω_b, and the barrier-height ΔV in the absence of the tilting. $\theta L (0 < \theta < 1)$ is the width of the left part of the barrier as in Fig.1. ω_R is the so-called

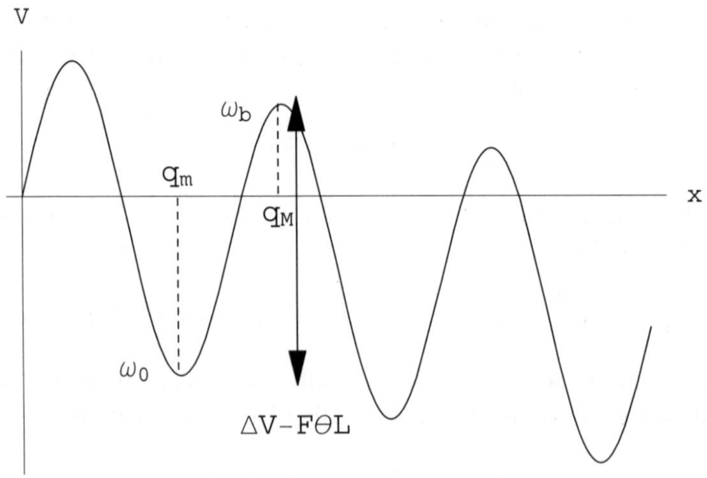

Fig. 1. Schematic illustration of the tilted periodic potential.

Grote-Hynse frequency.[26] Important prefactor f_q comes from the second-order variation of the action S_{eff} around the stationary orbits $q(\tau) = q_m$, and $q(\tau) = q_M$

$$f_q = \Pi_{n=1}^{\infty} \frac{\nu_n^2 + \omega_0^2 + \nu_n \hat{\gamma}(\nu_n)}{\nu_n^2 - \omega_b^2 + \nu_n \hat{\gamma}(\nu_n)}. \tag{6}$$

3. Local Detailed Balance for Quantum Thermal Hopping

Since the escape rate is now available, we can introduce a random walk on one-dimensional lattice which corresponds to the quantum Brownian motion in a tilted L-periodic potential. Each site of the lattice represents local minimum, i.e. well region. The transition probabilities to the nearest neighboring sites are proportional to the escape rates, $p_{+/-} \propto k_{+/-}$. By applying the periodic boundary condition, the distribution converges to a well-defined nonequilibrium steady state. Since the escape rate contains the quantum correction as a common factor f_q, the ratio between the escape rates k_+ and k_- towards positive and negative directions of the tilting satisfies the local detailed balance relation

$$\frac{k_+}{k_-} = e^{\beta F L}, \tag{7}$$

which is exactly the same relation as that of the classical tilted ratchet.[27]

4. Fluctuation Theorem

We assume that the entropy production ΔS is evaluated as the work done divided by the temperature. The probability density that the entropy production becomes $\beta F \Delta x$ with the net displacement $\Delta x = L \cdot k (k \in Z)$ is given as

$$\begin{aligned} & P(\Delta S = \beta F \Delta x) \\ &= \sum_{m=0}^{n} \sum_{k=-n}^{n-2m} \frac{n!}{(m+k)!m!(n-k-2m)!} p_+^{k+m} p_-^m (1-(p_+ + p_-))^{n-k-2m} \end{aligned} \tag{8}$$

In a short transient duration, initially localized distribution can give a negative net entropy production with a relatively large probability as in Fig.2.

The ratio between $P(\Delta S = \beta F \Delta x)$ and $P(\Delta S = -\beta F \Delta x)$ is easy to calculate, because the ratio between the generation probability of each orbit and its time reversal is simply $(\frac{p_+}{p_-})^k$, which is fully characterized by the net displacement $k = \frac{\Delta x}{L}$.Thus the fluctuation theorem is derived as

$$\frac{P(\Delta S = \beta F \Delta x)}{P(\Delta S = -\beta F \Delta x)} = e^{\beta F \Delta x}. \tag{9}$$

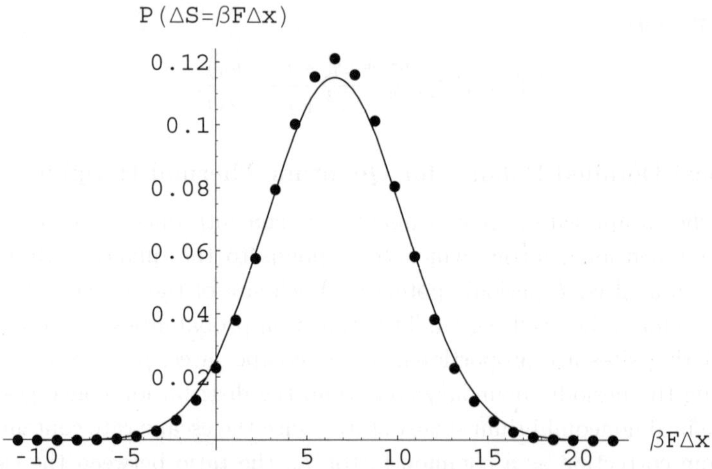

Fig. 2. Fluctuation of entropy production at a transient time. The array of dots is the distribution for the random walk (8). The Gaussian fitting is choosen so that the fluctuation theorem is satisfied (solidline). The value $\beta FL = 3$ is assumed.

It is remarkable that within the semiclassical escape rate treatment, we have the same symmetry as classical systems.

5. Large Deviation Expression

The fluctuation theorem is rewritten into the symmetry for generating function of the entropy production. Let us utilize the action functional of Lebowitz and Spohn,[3]

$$W(t) = \log \frac{k_{\sigma_{s1}}}{k^*_{\sigma_{s1}}} \frac{k_{\sigma_{s2}}}{k^*_{\sigma_{s2}}} \cdots \frac{k_{\sigma_{sn}}}{k^*_{\sigma_{sn}}}, \tag{10}$$

where the transitions occur at discrete times $0 = s_1 < s_2 < .. < s_n = t$ and $\sigma_t = +, -$ specifies whether the jump at time t is towards positive or negative directions of the perturbation, and the operation $*$ represents the time reversal, $k^*_{+/-} \equiv k_{-/+}$. Intuitively, the action functional represents the current for individual paths. Indeed, for the thermal average we have

$$\langle W(t) \rangle = \beta F J^{\rm st} t, \tag{11}$$

where $J^{\rm st} = L(k_+ - k_-)$ is the steady state current.The generating function $\mathrm{e}(\lambda)$ for the action functional $W(t)$ is given by,

$$\mathrm{e}(\lambda) = \lim_{t\to\infty} -\frac{1}{t}\log\langle e^{-\lambda W(t)}\rangle$$

$$= \lim_{n\to\infty} -\frac{1}{n\tau}\log\sum_{m=0}^{n}\sum_{k=-m}^{n-2m}$$

$$\left(\frac{p_+}{p_-}\right)^{k\lambda}\frac{n!}{(m+k)!m!(n-k-2m)!}p_+^{m+k}p_-^{m}\left(1-(p_+ + p_-)\right)^{n-k-2m}$$

$$= -\frac{1}{\tau}\log(1-(1-e^{-\lambda\beta FL})p_+ + (e^{\lambda\beta FL}-1)p_-). \tag{12}$$

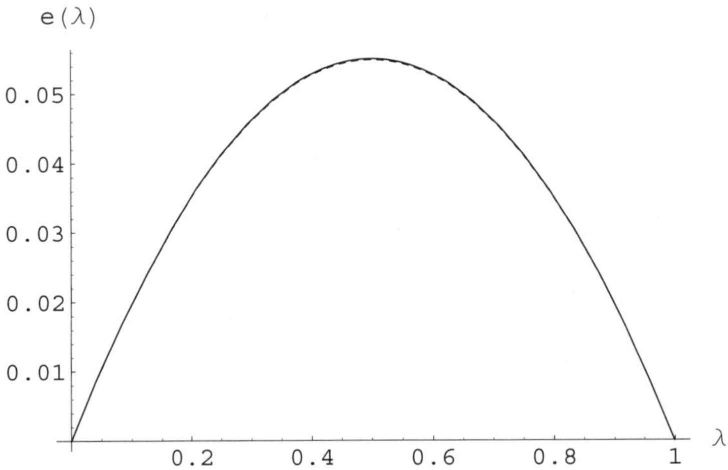

Fig. 3. Generating function $\mathrm{e}(\lambda)$ (12) (solidline). The broken-line (almost coincides with the solidline) shows the generating function $\beta FL(k_+ - k_-)\lambda(1-\lambda)$ of Gaussian distribution in Fig.2.

Noting that $\frac{p_+}{p_-} = \frac{k_+}{k_-} = e^{\beta FL}$, we have a symmetry

$$\mathrm{e}(\lambda) = \mathrm{e}(1-\lambda), \tag{13}$$

which is another expression of fluctuation theorem.

6. Summary

Fluctuation theorem is derived for a tilted periodic potential in the coexistence of the quantum tunneling and the thermal hopping for high enough temperature. In contrast to the operator formalism of quantum fluctuation theorems which deal with the energy of the total system, the escape rate formalism will be experimentally accessible by Josephson junctions. Indeed the phase diffusion can be detected as the decay of particular voltage state.[28] In this way, our fluctuation theorem might be experimentally accessible.

The author owes much to the 21st century COE program.

References

1. D.J. Evans, E.G.D. Cohen and G.P. Morriss, Phys. Rev. Lett.**71**, 2401(1993)
2. G.Gallavotti and E.G.D.Cohen, Phys.Rev.Lett.**74**,2694(1995)
3. J.L.Lebowitz, and H.Spohn, J.Stat.Phys.**95**333(1998)
4. C.Jarzynski,J.Stat.Phys.**98**, 77(2000)
5. G. M. Wang, E. M. Sevick, E. Mittag, D. J. Searles, and D. J. Evans, Phys. Rev. Lett.**89**, 050601 (2002)
6. D. M. Carberry et al., Phys.Rev.Lett.
7. D.Andrieux, and P.Gaspard, J.Stat.Phys.**127** 107(2007)
8. C.Jarzynski, Phys. Rev. Lett. **78**.2690.(1997)
9. G.E. Crooks, Phys. Rev. E **60**, 2721(1999)
10. G.E. Crooks, Phys.Rev.E.**61**, 2361-2366(2000)
11. W.Thirring, *Quantum Mathematical Physics; Atoms, Molecules and Large systems*, Springer, New York (2002)
12. J. Kurchan, e-print cond-mat/0007360v2
13. S. Tasaki, and T. Matsui, in: *Fundamental Aspects of Quantum Physics*, eds. L. Accardi and S. Tasaki, World Scientific, Singapore, pp.100. (2003)
14. T.Monnai, Nonlinear Phenomena in Complex Systems, **10** 102-104 (2007)
15. T.Monnai, and S.Tasaki, cond-mat/0308337
16. P.Talkner, E.Lutz, and P.Hanggi, Phys.Rev.E, **75**, 050102(R) (2007)
17. A.E.Allahverdyan and T.M.Nieuwenhuizen, Phys. Rev. E **71**, 066102 (2005)
18. S.Mukamel, cond-mat/0701003
19. J.Ankerhold, *Quantum Tunneling in Complex Systems The Semiclassical Approach*, Springer-Verlag Berlin (2007)
20. Ford, Lewis and O'Connell, Phys. Rev. A, **37** 4419 (1988)
21. C.W.Gardiner, IBM J.RES.DEVELOP.VOL.32 NO.1 (1988)
22. C.G.Callen, and S.Coleman, Phys.Rev.D, **16** 1762 (1977)
23. P.G.Wolynse, Phys.Rev.Lett.**47** 968 (1981)
24. H.Grabert, and U.Weiss, Phys.Rev.Lett.**53** 1787 (1984)
25. I.Affleck, Phys.Rev.Lett.**46** 388 (1981)
26. R.F.Grote, and J.Hynse, J.Chem.Phys.**73**, 2715(1980)
27. T.Monnai, J.Phys.A.Math.Gen.**37**, L75-79(2004)
28. L.D.Jackel, J.P.Gordon, E.L.Hu, R.E.Howard, L.A.Fetter, and J.Kurkijarvi, Phys.Rev.Lett.**47** 697 (1981)

STATISTICAL PROPERTIES OF THE INTER-OCCURENCE TIMES IN THE TWO-DIMENSIONAL STICK-SLIP MODEL OF EARTHQUAKES

T. HASUMI* and Y. AIZAWA

Department of Applied Physics, Advanced School of Science and Engineering, Waseda University, Tokyo 169-8555, Japan.
**E-mail: t-hasumi.1981@toki.waseda.jp*

We study earthquake interval time statistics, paying special attention to interoccurrence times in the two-dimensional (2D) stick-slip (block-slider) model. Interoccurrence times are the time interval between successive earthquakes on all faults in a region. We select stiffness and friction parameters as tunable parameters because these physical quantities are considered as essential factors in describing fault dynamics. It is found that interoccurrence time statistics depend on the parameters. Varying stiffness and friction parameters systematically, we optimize these parameters so as to reproduce the interoccurrence time statistics in natural seismicity. For an optimal case, earthquakes produced by the model obey the Gutenberg-Richter law, which states that the magnitude-frequency distribution exhibits the power law with an exponent approximately unity.

Keywords: Earthquakes; faulting; stick-slip model; interoccurrence time; self-organized phenomena.

1. Introduction

Earthquakes are caused by a fracture and frictional slip process. We can understand qualitatively how an earthquake occurs on the basis of the plate tectonic theory proposed by A. Wegener. Statistical properties of earthquakes are well known as empirical laws,[1] while the source mechanism of earthquakes is still an open question. For example, Gutenberg and Richter proposed the relation between magnitude (M) and frequency (n) expressed by

$$\log_{10} n = a - bM, \tag{1}$$

where a and b are positive constants. This relation is called the Gutenberg-Richter (GR) law.[2] b is the so-called b-value and similar to unity. Strictly speaking, b depends on fault structures and seismicity and ranges from 0.8 to 1.2.[3]

In general, earthquakes can be categorized into three types: foreshocks, mainshocks, and aftershocks. A mainshock is a large earthquake, whereas a foreshock (aftershock) is an earthquake before (after) the mainshock and which occurred near the mainshock epicenter. Aftershocks obey the Omori law,[4] which stresses that the decay rate of aftershocks follows the power law. Subsequently, a modified version was proposed by Utsu.[5] Since the Gutenberg-Richter law and the Omori law exhibit the power law, earthquakes are seemed to be self-organized critical phenomena.[6,7]

The time intervals between earthquakes can be classified into two types: recurrence times and interoccurrence times. Recurrence time is the interval of time between earthquakes on a single fault or segment, whereas interoccurrence time is the time interval between earthquakes on all faults in a region. For interoccurrence time statistics, probability distributions have been studied by different authors[8–10] by using earthquake catalogs (see fig. 1). Recurrence times are generally used by seismologists to describe the time interval between characteristic earthquakes.[11] A characteristic earthquake is a large earthquake happening on a single fault and depending on fault length, crust structure, and so on. For recurrence time, several probability distributions have been proposed, such as the log normal distribution, the Weibull distribution, and the double exponential distribution.[12] However, we cannot decide which distribution is the best owing to the lack of data.

In this work, we focus on the interoccurrence time statistics. Statistical properties based on earthquake models have been investigated numerically and compared with seismicity in nature. Then, earthquake models have been modified so as to reproduce fault systems.[13] Generally, numerical simulations have the advantage of enough earthquakes having occurred to guarantee statistical accuracy. In addition, it is easy to study the statistical properties under various crust conditions by changing control parameters. Optimizing or restricting control parameters so as to adequately reproduce the statistical properties, we suggest the probability distribution function of the recurrence time and offer new insights into earthquake statistics.

The stick-slip model proposed by Burridge and Knopoff[14] is often called the block-slider model or the Burridge-Knopoff model. This model describes the relative motion of faults. Although the model is highly simplified, it has been shown that it can extract the statistical properties of earthquakes, such as the GR law, the Omori law, the empirical law of the stress distribution, the constant stress drop, and the interoccurrence time statistics.[15–25] This

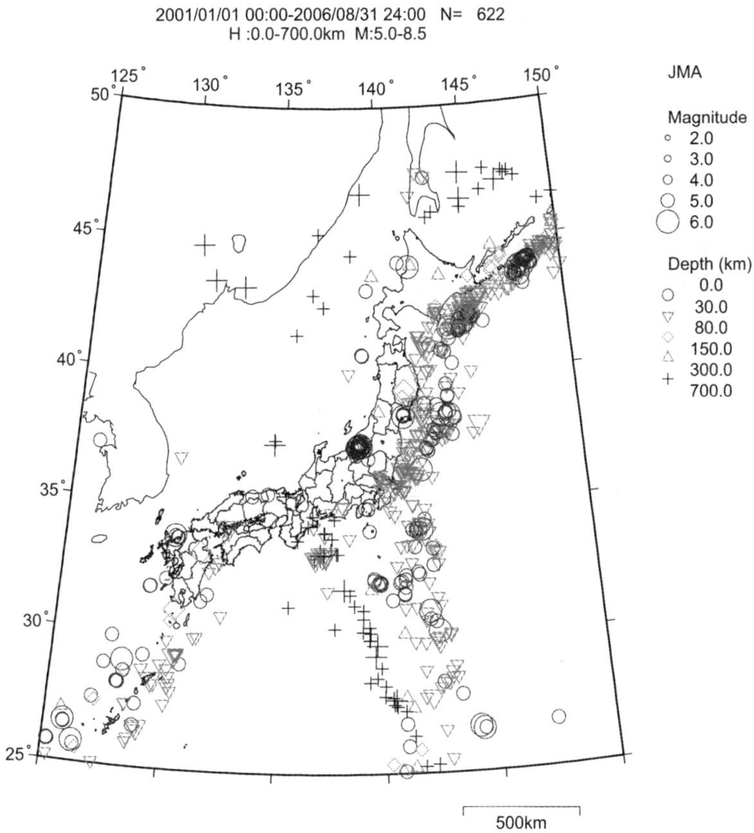

Fig. 1. Seismicity map around the Japan from 2001/01/01 to 2006/08/31 for $M > 5.0$. This map is programed on the basis of the JMA earthquake catalogs.

model has been modified in order to describe real crust structures.[26–28] However, the statistics of time intervals in the stick-slip model has not been discussed fully. The purpose of this work is to reveal whether the 2D stick-slip model can be understood as useful model in view of the inter-occurrence time statistics. Thus, we report numerical investigation of earthquake inter-occurrence time statistics produced by the 2D stick-slip model. In this work, stiffness and friction parameters are set as control parameters. Then we restrict or optimize these parameters so as to reproduce the inter-occurrence time statistics in nature. It is concluded that the model reproduces the inter-occurrence time statistics as well as the GR law in a limited parameter regime. Investigating the 2D stick-slip model in a optimal case further, we may propose new findings concerning the statistical properties of earthquakes.

2. Two-dimensional Stick-slip Model

In this work, we numerically investigate statistical properties of the interoccurrence time, produced by the two-dimensional (2D) stick-slip model. As shown in Fig. 2 (a), the model is composed of blocks on a square lattice, two plates, and two different kinds of springs. The upper plate is fixed, whereas the bottom plate moves at a velocity of v. So, this model represents the relative motion of faults. A block corresponds to a segment of a fault, so that we define a one-block-slip event as a minimum earthquake. Shear stresses and compression stress are modeled respectively by coil springs, k_c^x, k_c^y interconnected by a block, and by the leaf springs, k_p between a block and the fixed plate. We set the $-y$ axis as the direction of the loading plate, and the x axis as the perpendicular to the y axis. Assuming that the slip direction of a block is restricted by the y-direction only, the block is described by a stick-slip motion. The stick-slip motion can be divided in two parts: one is a stick-state and the other is a slip-state. In the case of the stick-state or loading-state, all blocks and the loading plate move together, whereas a block slips on the bottom plate, the slip state.

The equation of motion in a scaled form at cite (i, j) can be expressed by

$$\frac{d^2U_{i,j}}{dt'^2} = l_x^2(U_{i+1,j} + U_{i-1,j} - 2U_{i,j}) + l_y^2(U_{i,j-1} + U_{i,j+1} - 2U_{i,j}) - U_{i,j} - \phi\left(2\alpha\left(\nu + \frac{dU_{i,j}}{dt'}\right)\right), \tag{2}$$

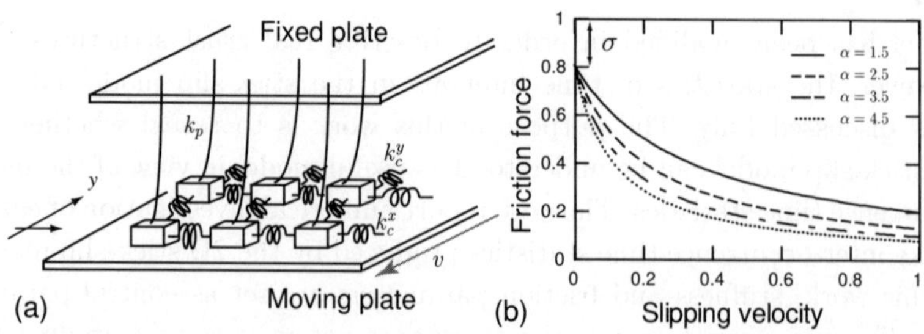

Fig. 2. (a) 2D stick-slip model. k_c^x, k_c^y, and k_p are spring constants. The friction force acts on the surface between the block and the bottom plate. (b) α-dependence of the non-linear dynamical friction function. σ is fixed at 0.01 throughout the simulation.

where U, t', and ϕ correspond respectively to a normalized displacement, time, and a dynamical friction force which is a function of block velocity. Additionally,

$$l_x^2 = \frac{k_c^x}{k_p}, \quad l_y^2 = \frac{k_c^y}{k_p}, \quad \nu = \frac{v}{\hat{v}}, \quad 2\alpha = \frac{\hat{v}}{v_1},$$

where $\hat{v}$ is a maximum of the slipping velocity and v_1 is the characteristic velocity. The blocks are subject to the friction acting on the surface between the block and the loading plate. In this study, we adopt "velocity-weakening" type friction law as a dynamical friction force, ϕ. This friction law states that a dynamical friction force decreases as the slipping velocity increases. This friction property can be observed in rock-fracture experiments[31] and is formulated mathematically by Carlson *et al.*,[17] namely,

$$\phi(\dot{U}) = \begin{cases} (-\infty, 1] & \dot{U} = 0, \\ \dfrac{(1-\sigma)}{\{1 + 2\alpha[\dot{U}/(1-\sigma)]\}} & \dot{U} > 0. \end{cases} \tag{3}$$

It is easy to simulate this friction law so that this formulation has been often used in previous works.[19,20,22,25,26,28] The friction function can be characterized by two parameters, σ and α. σ is the difference between the maximum friction force $(= 1)$ and the dynamical friction force at $v = 0$ $(= \phi(0))$. α is represented the decrement of the friction force, ϕ. If $\alpha = 0$, ϕ is constant, $1 - \sigma$. When $\alpha \to \infty$, ϕ decreases quickly to 0. ν is the normalized plate velocity and is very small parameter. Thus, we set $\nu = 0$ when an event happens. This assumption guarantees the condition that no other event occurs during an ongoing event.

In this study, we place 50×50 blocks on the (x, y) plane and simulate equation (2) and (3) under the free boundary condition by using the 4th-order Runge-Kutta method. Initial configurations of all blocks have small irregularities. 10^5 order of events after some periods when the initial randomness effect cannot be influenced are used. We study the interoccurrence time statistics by selecting l_x, l_y and α for tunable or control parameters, while $\nu = 0.01$ and $\sigma = 0.01$. This work is another version of the previous reports.[25]

3. Results and Discussion

In this work, the slip of a block is considered as an earthquake. An earthquake occurs when a block slips for the first time during an event. The interoccurrence time is defined as the time interval between successive events. For example, the nth interoccurrence time can be described by $\tau_n = t'_{n+1} - t'_n$, where t'_n and t'_{n+1} are the nth and the $n+1$th earthquake occurrence time, respectively. In order to compare our results with natural earthquake interoccurrence time, we introduce scaled interoccurrence time, $\tau' = \tau/\tilde{\tau}$, where $\tilde{\tau}$ is the normalized scaled time. $\tilde{\tau}$ is set at 1.0 for the model analysis and at 1000 [s] for observation analysis.

3.1. *Probability density distribution*

Probability density distributions of the interoccurrence time, $p(\tau')$ are displayed in Fig. 3. For (a), the friction parameter α is changed from 2.5 to 4.5, whereas stiffness parameter is fixed at $l_x = 1$ and $l_y = \sqrt{3}$. For (b), l_x and l_y are varied when $\alpha = 3.5$. As shown these figures, probability distributions of interoccurrence time exhibit the power law in short time scale region, $1 \lessapprox \tau \lessapprox 7$. We calculate the power law exponent, denoted here β from the slope of the distribution; for example, 2.40 ($\alpha = 2.5$), 1.96 ($\alpha = 3.5$), 1.78 ($\alpha = 4.5$) for (a). In the case of $\alpha = 3.5$ the distribution shows the power law, $1 \lessapprox \tau \lessapprox 20$, so that the system exhibits a critical state approximately (type A). For $\alpha = 2.5$, the probability of a long interoccurrence time region, $\tau \gtrapprox 10$, is enhanced more than expected by the power law decay, hereafter referred to as type B. On the other hand, as when $\alpha = 4.5$, the probability is less than predicted by the power law (type C). As for (b), β increases when l_x and l_y are enhanced, such as $\beta \simeq 1.94$, 1.96, and 2.71 for ($l_x = l_y = 1$), ($l_x = 1$ and $l_y = \sqrt{3}$), and ($l_x = 2$ and $l_y = 2\sqrt{2}$)qcThe forms and trends

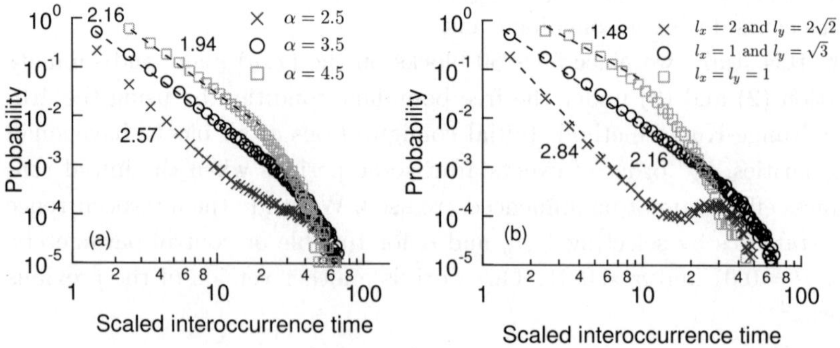

Fig. 3. The probability distributions of the inter-occurrence time for different control parameters, l_x, l_y, and α as a function of the scaled inter-occurrence time. For (a), $\times$ ($\alpha = 2.5$), $\circ$ ($\alpha = 3.5$), and $\square$ ($\alpha = 4.5$), while $l_x = 1$ and $l_y = \sqrt{3}$. For (b), $l_x = l_y = 1$, $l_x = 1$ and $l_y = \sqrt{3}$, and $l_x = 2$ and $l_y = 2\sqrt{2}$, denoted respectively, $\times$, $\circ$, and $\square$. b' is calculated from the slope of the dash line. All plots except for in the case of $\alpha = 3.5$ in (a) and $l_x = 1$ and $l_y = \sqrt{3}$ in (b) are shifted vertically for clarity.

of the distributions can be categorized into three types, type A ($l_x = 1$ and $l_y = \sqrt{3}$), type B ($l_x = 2$ and $l_y = 2\sqrt{2}$), and type C ($l_x = l_y = 1$). It is found that the form of the distribution in a long time region and the power law exponent β depend on the dynamical parameters, l_x, l_y, and α. Our findings are different from those of previous works studying of the 1D stick-slip model.[21–24]

3.2. *Survivor function*

Up to now, we have been discussing statistical properties of the survivor function, $D(\tau) = prob\ (t > \tau) = 1 - F(\tau)$, where $F(\tau)$ is the cumulative distribution. Abe and Suzuki have analyzed the Japan and Southern California earthquake catalogs and found that the survivor function of the interoccurrence time can be described by the power law as[10]

$$D(\tau') = \frac{1}{(1 + \epsilon\tau)^{\gamma}}, \tag{4}$$

where γ and $\bar{\tau}$ are parameters. This distribution function can be written as

$$D(\tau') = e_q(-\tau/\tau_0) = [(1 + (1-q)(-\tau/\tau_0))^{\frac{1}{1-q}}]_+,\ ([a]_+ \equiv \max[0, a]), \tag{5}$$

where q and τ_0 are positive constants and are related to γ and ϵ: $\gamma = 1/(q-1)$ and $\epsilon = (q-1)/\tau_0$. $e_q(x)$ is the so-called q-exponential distribution derived from the non-additive statistical mechanics proposed by Tsallis.[30] In this work, we select the power law distribution defined by

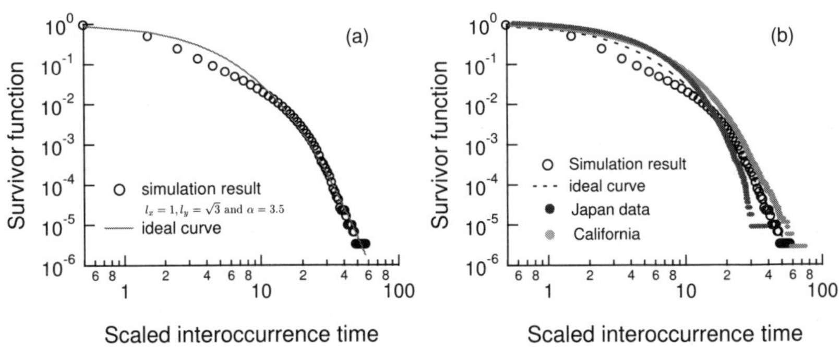

Fig. 4. The survivor functions of inter-occurrence time statistics. For (a), the simulation is performed $l_x = 1, l_y = \sqrt{3}$, and $\alpha = 3.5$. For (b) the survivor distributions for the optimal case of the model, Japan earthquakes, and California earthquakes are shown. The fitting parameters are calculated as $q = 1.06$ and $\tau_0 = 2.55$ for the model, $q = 1.05$ and $\tau_0 = 3.16$ for Japan, and $q = 1.13$ and $\tau_0 = 3.44$ for California.

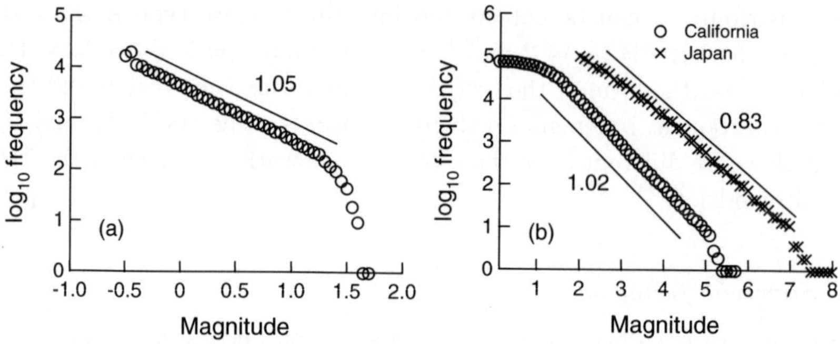

Fig. 5. The magnitude distribution obtained from the model in (a) and from real earthquakes in (b). The b-value of the GR law is calculated from the slope of the distribution: 1.05 (model), 0.83 (Japan), and 1.02 (California).

Eq. (5) for the ideal survivor distribution function, and then optimize the control parameters, l_x, l_y and α.

We present the survivor function of the interoccurrence time in the case of $l_x = 1, l_y = \sqrt{3}$, and $\alpha = 3.5$ in Fig. 4 (a). Plots and the dashed line correspond to the numerical data and the ideal curve, respectively. Varying the control parameters, l_x, l_y, and α systematically, we can find the optimal case of the control parameters: $l_x = 1, l_y = \sqrt{3}$, and $\alpha = 3.5$. For an optimal case, the fitting parameters of the survivor functions are estimated to be $q = 1.06, \tau_0 = 2.55$ and the correlation function ρ yields 0.986. Figure 4 (b) shows the interoccurrence time statistics obtained from the model (optimal case), Japan earthquakes, and California earthquakes. Therefore, it is concluded that the interoccurrence time can be reproduced semi-qualitatively for the optimal case.

Finally, magnitude distributions derived from the model in the case for $l_x = 1, l_y = \sqrt{3}$, and $\alpha = 3.5$ and from Japan and California earthquake catalogs are shown in Fig. 5 (a) and (b), respectively. Note that we use the JMA catalogs: "http://kea.eri.u-tokyo.ac.jp/tseis/jma1/" for Japan and the NCEDC catalogs: "http://www.ncedc.org/ncedc/catalog-search.html" for California for the period 2001/01/01-2006/08/31. It should be noted that the JMA catalog lacks earthquake data whose magnitude is less than 2.0. Comparing Fig. 5 (a) with (b), we demonstrate that the model can reproduce the GR law, which we explained previously: the power law distribution with exponent, b-value 1.0. We found that in the case of optimal parameters, the interoccurrence time statistics and the GR law can be extracted from the 2D stick-slip model.[25]

4. Conclusion

In this study we numerically investigated the statistics of earthquake interval times, the interoccurrence time based on the 2D spring-block model. Interoccurrence times are interval times between earthquakes on all faults in a region. It is found that interoccurrence time statistics depend on the control parameters, l_x, l_y, and α charactering the model. l_x and l_y are stiffness parameters, whereas α expresses the decrement of dynamical friction. The probability density distributions of the interoccurrence time show the power law in the short-time region. For the long time region, the distributions could be classified into three types: power law behavior (type A), broad peak structures (type B), and exponential cutoffs (type C). Then, we restricted the control parameters so as to reproduce the interoccurrence time statistics in nature; the survival function of the interoccurrence time revealed the q-exponential distribution with $q > 1$. The optimal parameters are estimate to be $l_x = 1, l_y = \sqrt{3}$, and $\alpha = 3.5$.

In the case of the optimal parameters, the magnitude distribution shows the power law with exponent 1.0. This power law distribution is similar to the GR law, and the exponent corresponds to the b-value, which is characterized by the GR law. Hence, we demonstrate that in the optimal case, $l_x = 1, l_y = \sqrt{3}$, and $\alpha = 3.5$, the model can reproduce the GR law and the interoccurrence time statistics both simultaneously and spontaneously.

We acknowledge that the stick-slip (block-slider) model is highly simplified so that many effects playing important roles in fault dynamics have been neglected. However, it is shown that this model is useful for the extraction of the statistical properties of earthquakes because the interoccurrence time statistics and the GR law can be extracted. This work is a first step toward studying the origin of the statistics of time intervals. We hope to extend our work by focusing on recurrence time statistics and comparing them with natural recurrence time statistics.

Acknowledgments

T.H. would like to thank Prof. S. Abe, Prof. N. Suzuki, Prof. H. Kawamura, Dr. M. Kamogawa, Dr. T. Sato, Dr. T. Hatano, Dr. Y. Kawada, and Dr. T. Mori for useful comments, fruitful discussions, and manuscript improvements. This work was partly supported by the Japan Society for the Promotion of Science (JSPS), the Earthquake Research Institute cooperative research program at University of Tokyo, and a grant to the 21st Century COE Program "Holistic Research and Education Center for Physics of Self-organization Systems" at Waseda University from the Ministry of Education, Culture, Sports, Science, and Technology (MEXT), Japan.

References

1. I. G. Main, *Rev. Geophys.*, **34**, 433 (1996).
2. B. Gutenberg and C. F. Richter, *Ann. Geophys.*, **9**, 1 (1956).
3. C. Frohlich and S. D. Davis, *J. Geophys. Res.*, **98**, 631 (1993).
4. F. Omori, *J. College Sci., Imp. Univ. Tokyo*, **7**, 111 (1894).
5. T. Utsu, *Geophys. Mag.*, **30**, 521, (1961).
6. P. Bak, C. Tang, and K. Wiesenfeld, *Phys. Rev. A.*, **38**, 364 (1988).
7. P. Bak, and C. Tang, *J. Geophys. Res.*, **94**, 15635 (1989).
8. P. Bak, K. Christensen, L. Danon and T. Scanlon, *Phys. Rev. Lett.*, **88**, 178501 (2002).
9. A. Corral, *Phys. Rev. Lett.*, **92**, 108501 (2004).
10. S. Abe and N. Suzuki, *Physica A*, **350**, 588 (2005).
11. M. W. String, S. G. Wesnousky and K. Shimazaki, *Geophys. J. Int.*, **124**, 833 (1996).
12. M. V. Matthews, W. L. Ellsworth and P. A. Reasenberg, *Bull. Seism. Soc. Am.*, **92**, 2233 (2002).
13. J. B. Rundle, D. L. Turcotte, and W. Klein, *GeoComplexity and the Physics of Earthquakes*, (American Geophysical Union, Washington, DC, 2000).
14. R. Burridge and L. Knopoff, *Bull. Seism. Soc. Am.*, **57**, 341 (1967).
15. J. M. Carlson and J. S. Langer, *Phys. Rev. Lett.*, **62**, 2632 (1989).
16. J. M. Carlson and J. S. Langer, *Phys. Rev. A*, **40**, 6470 (1989).
17. J. M. Carlson, J. S. Langer, B. E. Shaw, and C. Tang, *Phys. Rev. A*, **44**, 884 (1991).
18. J. M. Calson and J. S. Langer, *J. Geophys. Res.*, **96**, 4255 (1991).
19. J. M. Carlson, *Phys. Rev. A*, **44**, 6226 (1991).
20. H. Kumagai, Y. Fukao, S. Watanabe and Y. Baba, *Geophys. Res. Lett.*, **26**, 2817 (1999).
21. E. F. Preston, J. S. Sa Martins, J. B. Rundle, M. Anghel, and W. Klein, *Comput. Sci. and Eng.*, **2**, 34, (2000).
22. T. Mori and H. Kawamura, *J. Geophys. Res.*, **111**, B07302 (2006).
23. A. Omura and H. Kawamura, *Europhys. Lett.*, **77**, 69001 (2007).
24. S. G. Abaimov, D. L. Turcotte, R. Shcherbakov, and J. B. Rudle, *Nonlin. Processes Geophys.*, **14**, 4551 (2007).
25. T. Hasumi, *Phys. Rev. E*, **76**, 026117 (2007).
26. H. Nakanishi, *Phys. Rev. A*, **46**, 4689 (1992).
27. S. Hainzl, G. Zöller and J. Kurths, *J. Geophys. Res.*, **104**, 7243 (1999).
28. J. Xia, H. Gould, W. Klein, and J. B. Rundle, *Phys. Rev. Lett.*, **95**, 248501 (2005).
29. Y. Yamashita, *J. Phys. Earth.*, **24**, 417, (1976).
30. C. Tsallis, *J. Stat. Phys.*, **52**, 479 (1988).
31. C. H. Scholz, *The Mechanics of Earthquakes and Faulting*, (Cambridge Univ Press, Cambridge, England, 2002).

SECOND HARMONIC GENERATION AND POLARIZATION MICROSCOPE OBSERVATIONS OF QUANTUM RELAXOR LITHIUM DOPED POTASIUM TANTALATE

HIROKO YOKOTA
YOSHIAKI UESU

Department of Physics, Waseda University, 3-4-1 Okubo Shinjuku-ku, Tokyo,169-8555, JAPAN

Polar state in a quantum relaxor $K_{(1-x)}Li_xTaO_3$ (KLT) is investigated using second harmonic generation (SHG) and polarization microscopes. Temperature dependences of SHG image and interference color image related to birefringence are observed on three different processes (zero field heating after zero field cooling, field heating after zero field cooling, and field heating after field cooling processes). A remarkable history dependence in the *T-E* space which is one of the characteristic behaviors of relaxor is observed in SHG and polarization microscope observations. Ferroelectric phase transition occurs below the transition temperature T_p with tiny domain structures which is beyond the optical microscope diffraction limit. Under an electric field, these micro domain structures change to macroscopic structures. Based on these experiments, we propose a polar state model of KLT below T_p.

1. Introduction

Relaxors are one of the most attractive ferroelectric-related materials from both points of view of applications and fundamental physics because of their colossal dielectric and piezoelectric responses. Relaxors are commonly defined by the following four criteria [1]; (i) characteristic dielectric dispersion in the low frequency region, [2] (ii) history dependence of the order parameter under an electric field, [3-9] (iii) slow kinetics of the order parameter under an electric field, [10-12] and (iv) existence of polar nano region (PNR). [13-19] After the discovery of so-called prototype relaxor $Pb(Mg_{1/3}Nb_{2/3})O_3$, [20,21] enormous researches have been performed. In spite of their efforts, the origin of relaxor behaviors has not been clearly understood up to now. However, it is basically accepted that the intrinsic heterogeneity in the system causes the peculiar behaviors.

$KTaO_3$ (KTO) is known to be one of the prototype quantum paraelectrics. [22] In quantum paraelectrics, the ferroelectric phase transition does not appear down to 0 K although the dielectric constant increases monotonically towards the low temperature. With doping Li ions, they are replaced by K ions and shift by 1 A from the ideal A-site positions. We have disclosed recently that KTO doped with Li (KLT) satisfies all four criteria of the relaxor nature by second harmonic generation (SHG) microscope observations, [23,24] X-ray diffraction, neutron scattering, [25] and dielectric measurements. [26] Based upon the fact that the quantum paraelectric KTO changes to relaxor with Li doping, we name KLT quantum relaxor. Furthermore we found that the large dielectric constant of KLT with the tri-critical point composition is originated from its location in the vicinity of the line of critical end points. In the present paper, we report the results of SHG and polarization microscope observations to elucidate the complicate polar state below the ferroelectric phase transition temperature T_p.

2. Experimental Conditions

2.1. *SHG microscope observations*

For the measurements, KLT with Li concentration of 2.6% (KLT-2.6%) is used as a specimen, because this concentration is located near the tri-critical point and also in the cross-over region from the quantum paraelectric to relaxor/ferroelectric state. A (100) plate sample with the dimension of 5x9.4x0.854 mm^3 is cut from a flux grown single crystal. The surfaces of the sample are optically polished with oxide aluminum. To apply a homogeneous electric field in the whole area, it is better to use transparent electrodes like ITO on both surfaces. However, a remarkable photocurrent is reported in KLT in low temperatures which is not negligible for the SHG microscope observation. Thus we evaporate gold electrodes apart by 3 mm at the top surface and a special care is paid not to illuminate the laser at the electrodes.

We use an Nd^{3+}:YAG laser with a wavelength of 1064 nm, repetition frequency of 20 Hz, the light intensity per pulse is 80 mW/cm^2. The illumination of pulsed laser beam does not produce a noticeable increase of the sample temperature and it is examined experimentally to be less than 0.6 K. The detailed optical system is described in ref. 23. The magnitude of applied electric field is 80 V/mm whose direction is parallel to the [001] in the surface plane.

The temperature dependences of SH intensity are measured in three different processes; (i) zero field heating after zero field cooling process

(ZFH/ZFC), (ii) field heating after ZFC process (FH/ZFC), and (iii) field heating after field cooling process (FH/FC).

2.2. *Polarization microscope observations*

A same specimen for the SHG measurement of KLT-2.6% is used for the polarization microscope observations. For applying an electric field, ITO transparent electrodes are put on the both surfaces by using a pulsed laser deposition method. To compare the results of SHG observations, we make measurements in the three processes; (i) ZFH/ZFC, (ii) FH/ZFC, and (iii) FH/FC. All measurements are performed under the cross Nicole condition: the directions of a polarizer and an analyzer are perpendicular to each other and the optical principal axes of the sample make an angle of 45 degree from the cross Nicole position. A color sensitive plate is inserted between a sample and an analyzer to detect small appearance of a birefringence.

3. Experimental Results and Discussions

3.1. *SHG microscope observations*

Temperature dependences of SH intensity in different processes are shown in Fig. 1. In ZFH/ZFC process, an almost null SH wave is generated in the sample in the whole temperature regions.

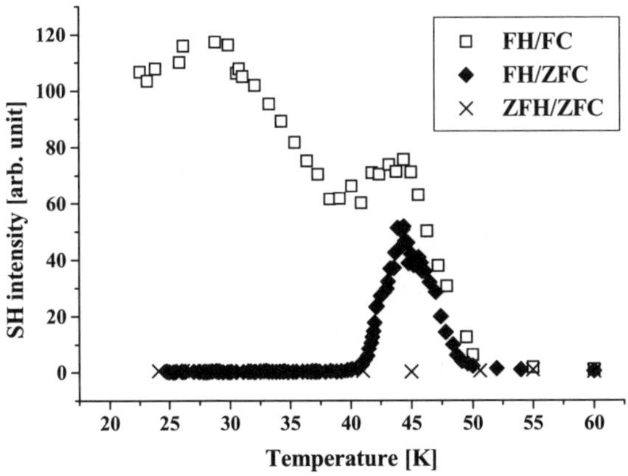

Fig. 1. Temperature dependences of SH intensity at different processes on KLT-2.6%.

In FH/ZFC process, SH intensity does not increase at low temperatures after applying an electric field. Corresponding SHG images are shown in Fig.3 (a).

With raising temperature, a weak SH intensity generated at a low temperature begins to increase around 40K. It takes a maximum value at 45 K and vanishes around 50K of T_p. On the contrary, strong SH waves which are produced at low temperature, decrease monotonically with heating the specimen and vanish around T_p in FH/FC process. It denotes that T_p is the temperature where the ferroelectric phase transition occurs.

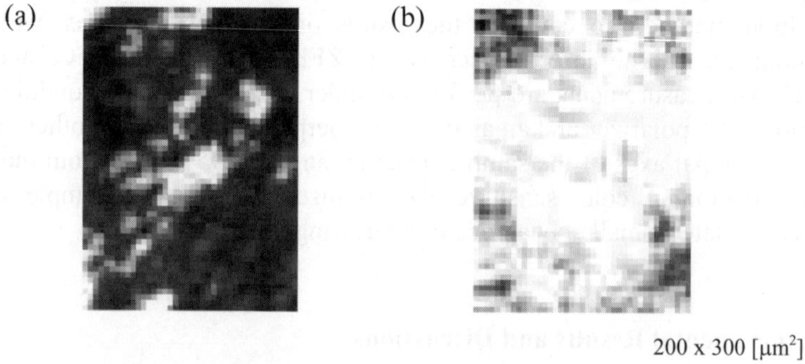

Fig. 2. SHG images from samples.
(a) shows the result of KLT-2.6% at 24K on FC process. (b) indicates that of $LiNbO_3$ at room temperature. Bright parts correspond to the regions generating strong SHG.

From these experiments, the history dependence of the order parameter is confirmed in KLT-2.6%, which is one of the four criteria of relaxor as mentioned above. In the relaxor system, the origin of their specific behaviors is interpreted to originate from the intrinsic heterogeneity. In Fig.2 (a), the SHG image of FH/FC process in KLT-2.6% is shown. Fig.2 (a) shows the coexistence of two regions, i.e., the regions which generate strong SH waves and the regions which do not generate them under an electric field, reflecting the intrinsic inhomogeneity of KLT. To confirm the homogeneity of Li concentration distribution, we choose several different regions to measure the SH intensities and verify that SH intensities from the different regions disappear almost the same temperature of T_p. It indicates that the inhomogeneity observed in Fig.2 does not come from the Li concentration gradient. To examine the inhomogeneity of the SH intensity distribution, we take an SHG image of single domain crystal $LiNbO_3$ which generates homogeneous SH waves at room temperature. The result is shown in Fig.2 (b). It should be noted that SH waves distribute with intensity fluctuation of 25 % from the $LiNbO_3$ sample, while they distribute with 75 % in the case of KLT-2.6%. The former fluctuation is caused by a quite small change of $\Delta n = n^{(2\omega)}-n^{(\omega)}$: in the present case of 25 %

fluctuation corresponds to the change of Δn of 1/6000. From these results, we conclude that the inhomogenity SH wave distribution reflects the intrinsic heterogeneity in KLT.

3.2. *Polarization microscope observations*

In ZFH/ZFC process, the domain structure does not appear even at low temperatures below T_p. On the other hand, the change in interference color is observed below T_p. This indicates that a macroscopic strain corresponding to the results of X-ray diffraction experiments appears below T_p.

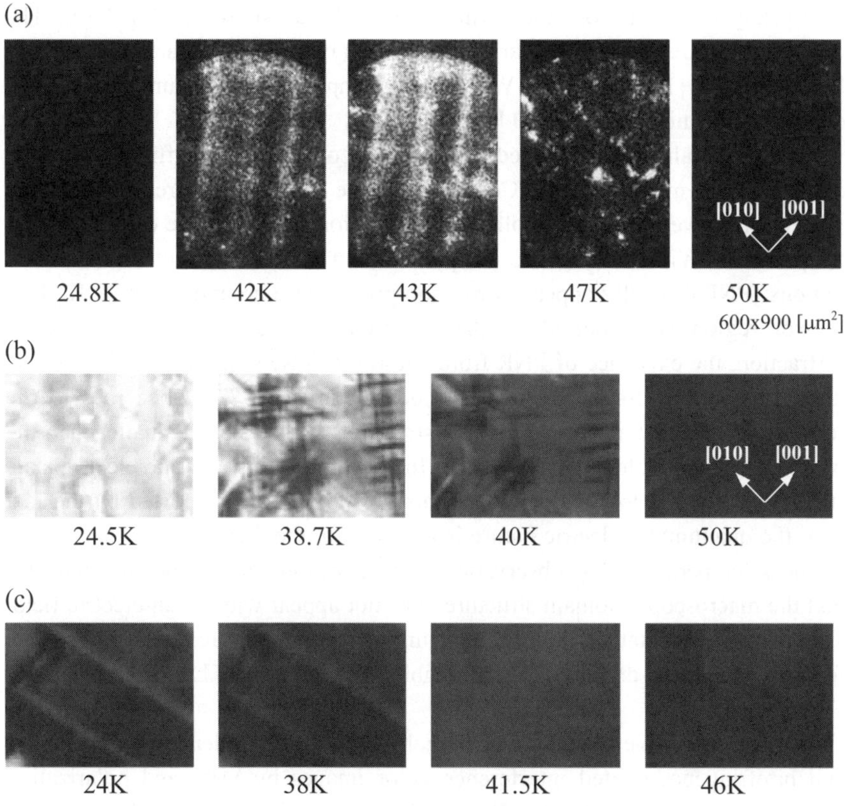

Fig. 3. Temperature dependence of SHG (a) and polarization microscope images on FH/ZFC (b) and those in FC (c) processes. In (a), bright parts correspond to the regions generating strong SHG.

Judging from these experimental results, i. e., no macroscopic domain structure characterizing the m-3mF4mm transition and almost zero SH intensity are observed without an electric field, we conclude that the size of ferroelectric domain is quite small and less than 1 micron.

In Fig.3 (b), the results of FH/ZFC process are shown in comparison with those of SHG microscope observation. At low temperatures, the interference color pattern is almost same with that of ZFH/ZFC process. With heating the sample up to 40 K, domain walls or domain structures appear along the [100] and [010] directions. This result is consistent with the SHG microscope observation: micro domains change to macro domains above 40 K. Around 50K, the interference color disappears in the whole region, which means that the sample is in the isotropic cubic phase. In FH/FC process, the behavior is completely different from the former two paths as shown in Fig.3 (c). The domain structure whose boundary plane is the (100) can be clearly observed as lines along the [100] direction. With raising temperature, the boundary becomes obscure and vanishes around 50 K.

We have already performed SHG microscope, neutron diffuse scattering, and dielectric measurements. Combining these experimental results with the present ones, we propose the following model for the polar state of KLT at low temperature. With cooling the sample without an electric field, polar nano regions (PNRs) locally appear in the non-polar phase around T_d of 90 K. This relaxor region is supported by the occurrence of tetragonality from X-ray diffraction, the existence of PNR from neutron diffuse scattering, and dielectric dispersion at low frequency region. Cooling down to T_p, it becomes to the ferroelectric phase with micro domain structures whose size is smaller than the transverse coherent length of the laser. In addition to this, the paraelectric state still remains at the low temperature. This presumption is supported by the fact that the quantum paralectric nature is observed in the dielectric constant below the peak temperature. The observation of polarization microscope also indicates that the macroscopic domain structure does not appear without an electric field. The paraelectric state which is too tiny to detect still remains among the ferroelectric micro domains. The probable polar state of KLT is illustrated in Fig.4.

In summary, we observe a remarkable history dependence of SH images and birefringence-related interference color images by SHG and polarization microscopes, respectively. In ZFH/ZFC process, almost null SH waves are generated and no domain structure is observed. However, a clear change of interference color appears below T_p. Thus, we conclude that below T_p it changes to the ferroelectric micro domain whose size is less than 1μm without an electric field.

(a) $E = 0$ (b) $E \neq 0$

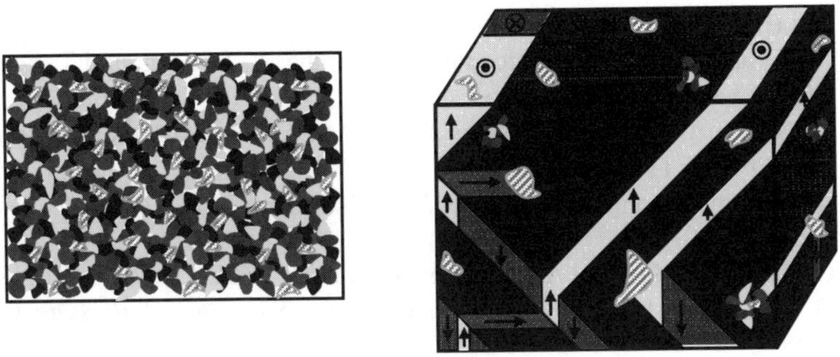

Fig. 4. Polar state model without E (a) and without E (b) below T_P.
In Fig.4(a), hatched regions indicate paraelectric phase, and other regions ferroelectric phase with different directions of Ps. In Fig.4(b), arrows indicate the direction of Ps, and hatched regions paraelectric phase.

References

1. L. E. Cross, *Ferroelectrics* **76**, 241 (1987).
2. A. J. Bell, *J. Phys.: Condens. Matter* **5**, 8773 (1993).
3. S. B. Vakchrushev, B. E. Kvyatkovsky, A. A. Nabereznov, N. M. Okuneva, and B. P. Toperverg, *Physica* **B156&157**, 90 (1989).
4. N. de Marhan, E. Husson, G. Calvarin, and A. Morell, *Mat. Res. Bull.* **26**, 1167 (1991).
5. V. Westphal, W. Kleemann, and M. D. Glinchuk, *Phys. Rev. Lett.* **68**, 847 (1992).
6. R. Sommer, N. K. Yushin, and J. J. van der Klink, *Phys. Rev.* **B48**, 13230 (1993).
7. O Bidault, M Licheron, E Husson and A Morell, *J. Phys.: Condens. Matter* **8,** 8017 (1996).
8. E. V. Colla, N. K. Yushin, and D. Viehland, *J. Appl. Phys.* **83**, 3298 (1998).
9. C. Ang, Z. Yu, P. Lunkenheimer, J. Hemberger, and A. Loidl, *Phys. Rev.* **B59**, 6670 (1999).
10. K. Fujishiro, T. Iwase, Y. Uesu, Y Yamada, B. Dkhil, J. M. Kiat, S. Mori, and N. Yamamoto, *J. Phys. Soc. Jpn.* **69**, 2331 (2000).
11. E. V. Colla, E. Yu. Koroleva, N. M. Okuneva, and S. B. Vakhrushev, *Phys. Rev. Lett.* **74**, 1681 (1995).
12. E. V. Colla, E. Yu. Koroleva, N. M. Okuneva, and S. B. Vakhrushev, *Ferroelectrics* **184**, 209 (1996).
13. B. Dkhil, and J. M. Kiat, *J. Appl. Phys.* **90**, 4676 (2001).
14. G. Burns, and F. H. Dacol, *Solid State Commun.* **48**, 853 (1983).

15. G. Burns, and F. H. Dacol, *Phys. Rev.* **B28**, 2527 (1983).
16. S. B. Vakhrushev, B. E. Kvyatkovsky, A. A. Nabereznov, N. M. Okuneva, and B. P. Toperverg, *Ferroelectrics* **90**, 173 (1989).
17. S. B. Vakhrushev, A. Nabereznov, S. K. Sinha, Y. P. Feng, and T. Egami, *J. Phys. Chem. Solids* **57**, (1996).
18. A. Nabereznov, S. Vakhrushev, B. Dorner, D. Strauch, and H. Moudden, *Eur. Phys. J.* **11**, 13 (1999).
19. D. La-Orauttapong, J. Toulouse, J. L. Robertson, and Z. –G. Ye, *ibid.* **64**, 212101 (2001).
20. G. A. Smolenskii, and A. I. Agranovskaya, *Zh. Thek. Fiz.* **28**, 1491 (1959).
21. G. A. Smolenskii, and A. I. Agranovskaya, *Sov. Phys. Sol. State* **1**, 1429 (1959).
22. S. H. Wemple, *Phys. Rev.* **137**, A1575 (1964).
23. H. Yokota, T. Oyama, and Y. Uesu, *Phys. Rev.* **B72**, 144103 (2005).
24. H. Yokota, and Y. Uesu, *J. Phys.: Condens. Matter* **19**, 102201 (2007).
25. H. Yokota, Y. Uesu, C. Malibert, and J. M. Kiat, *Phys. Rev.* **B75**, 184113 (2007).
26. H. Yokota, A. Okada, I. Ishida, and Y. Uesu, *J. Jpn. Appl. Phys.* **46**, 10B, 7167 (2007).

THERMOELECTRIC PROPERTIES OF Ni-DOPED $LaRhO_3$

S. SHIBASAKI*, Y. TAKAHASHI and I. TERASAKI

Department of Applied Physics, Waseda University, Tokyo, 169-8555, Japan
** E-mail: g01k0374@suou.waseda.jp*
www.f.waseda.jp/terra/

We report resistivity and thermopower of polycrystalline samples of Ni-doped $LaRhO_3$ from room temperature down to 4.2 K. The resistivity decreases with increasing Ni-doping, while the thermopower reaches almost a constant value ($\sim$85 μV/K) at room temperature.

Keywords: Thermoelectric properties; resistivity; thermopower; $LaRhO_3$.

1. Introduction

Oxides have potential to be good thermoelectric materials, especially at high temperatures thanks to their stability against oxidation. Thermoelectric properties of materials are evaluated by $ZT = S^2T/\rho\kappa$, where S, T, ρ, κ represent thermopower, absolute temperature, resistivity and thermal conductivity, respectively. To obtain large ZT, we need materials with large thermopower, low resistivity and low thermal conductivity.

Layered Co oxide Na_xCoO_2 has a thermopower as large as 100 μV/K with a low resistivity of 200 $\mu\Omega$cm at room temperature.[1] After the discovery of these good thermoelectric properties, other layered thermoelectric Co oxides, such as Bi-Sr-Co-O and Ca-Co-O, have been discovered.[2–4] Koshibae *et al.* have revealed that conduction between $Co^{3+/4+}$ ions in the low-spin state causes large thermopower at high temperatures.[5] Rh lies just below Co in the periodic table, and chemical properties are expected to be similar. Furthermore, Rh ions are stable in the low-spin state up to high temperature, and thus Rh oxides can be better thermoelectric materials than Co oxides at that temperature range. Rh oxides have been investigated by several researchers from this point of view.[6–11]

Perovskite Co oxide $LaCoO_3$ with a three-dimensional CoO_6 network shows good thermoelectric properties around room temperature.[12] The spin state of Co ions in $LaCoO_3$ changes from the low-spin to high/intermediate-spin state above 500 K at which the thermopower suddenly decreases.[12,13] In order to prevent this spin-state transition, we study an isomorphic Rh oxide, $LaRhO_3$. Here, we show the thermoelectric properties of Ni-doped $LaRhO_3$.

Figure 1 shows the crystal structure of $LaRhO_3$ which crystallizes in the $GdFeO_3$-type perovskite structure.[14] The RhO_6 octahedrons have edge-shared network like the CoO_6 octahedrons in $LaCoO_3$, which is responsible for electrical conduction. By the substitution of Ni^{2+} for Rh^{3+}, formal valence of Rh changes to $Rh^{(3+x)+}$ where x corresponds to the Ni content, and good conduction with large thermopower is expected like in Co oxides.[12]

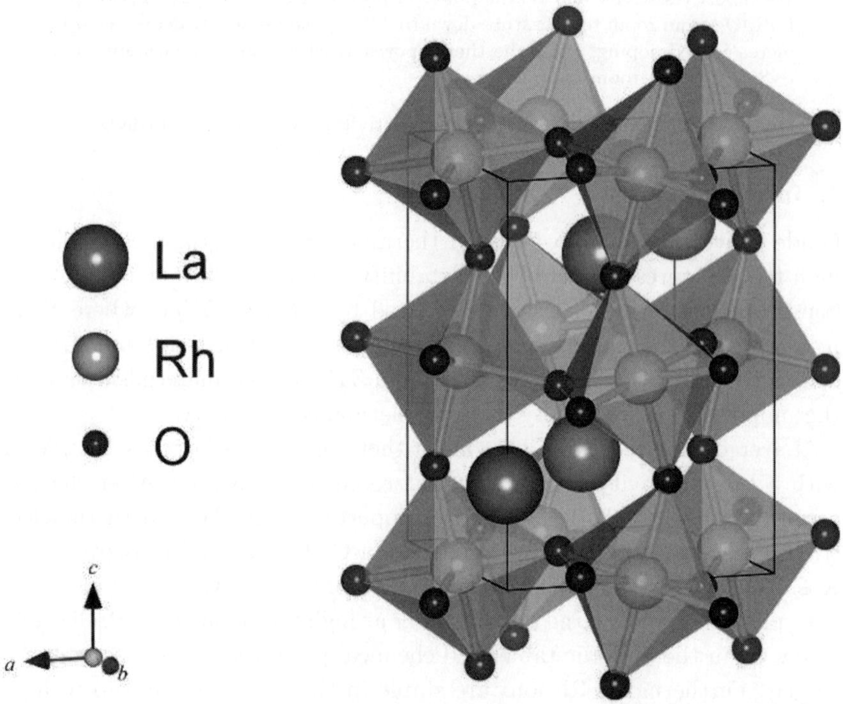

Fig. 1. The crystal structure of $LaRhO_3$. Large, medium and small spheres represent La, Rh and O atoms, respectively. This material has edge-shared RhO_6 octahedron network.

2. Experimental Method

Polycrystalline samples of Ni-doped $LaRhO_3$ were prepared by a solid-state reaction. Stoichiometric amounts of La_2O_3, Rh_2O_3 and NiO of 99.9% purity were mixed and calcined at 1000 °C for 1 day in air. The mixtures were thoroughly mixed, pelletized and sintered at 1100 °C for 2 days. The X-ray diffraction (XRD) patterns of the samples were obtained using a CuKα radiation by a θ–2θ method from 10 to 100 degree. Resistivity and thermopower were measured in a liquid He cryostat from 4.2 to 300 K. Resistivity was measured by a conventional dc four-probe technique, and thermopower was measured by a steady-state technique with a typical temperature gradient of 0.5 – 1 °C.

3. Results and Discussion

Figure 2 shows the XRD patterns of $LaRh_{1-x}Ni_xO_3$. We can see a tiny peak of NiO impurity only at $x = 0.3$ near $2\theta = 43$ deg. This means that our samples are almost in a single phase, and good enough to measure thermoelectric properties.

Figure 3 shows the measured resistivity and thermopower. The resistivity of $LaRhO_3$ is consistent with the previous report of Nakamura *et al.* (45 Ωcm at 300 K).[15] The resistivity decreases with the increase of Ni content down to 35 mΩcm for $x = 0.3$. This value is about 6 times larger than that of $LaCo_{0.7}Ni_{0.3}O_3$.[16] This is due to the larger distance between Rh and O atoms or a smaller tilting angle between adjacent RhO_6 octahedra ($\sim 149°$). The thermopower of $LaRhO_3$ is about 400 μV/K at room temperature, which is as large as that of $LaCoO_3$. With increasing Ni content, the thermopower decreases and reaches 80 μV/K at room temperature for $x = 0.3$ which is twice as large as that of $LaCo_{0.7}Ni_{0.3}O_3$.[16] In heavily-doped samples ($x \geq 0.10$), the thermopower is nearly independent of x, while it monotonically decreases with x in $LaCo_{1-x}Ni_xO_3$.[12] This tendency of $LaRh_{1-x}Ni_xO_3$ looks similar to the results of delafossite Rh oxide $Cu_{1-x}Ag_xRh_{1-y}Mg_yO_2$ and spinel Rh oxide $Zn(Rh_{1-x}Mg_x)_2O_4$.[10,11] We ascribed this thermopower to an electronic phase separation.

Table 1 lists the comparisons of the thermoelectric parameters of $LaRh_{0.7}Ni_{0.3}O_3$ and $LaCo_{0.7}Ni_{0.3}O_3$. The thermoelectric properties of Ni-doped $LaRhO_3$ are comparable to those of $LaCoO_3$ near room temperature. Above 500 K, the thermopower of Ni-doped $LaRhO_3$ does not decrease presumably due to the absence of the spin-state transition with increasing temperature.[17] In this point, we can say that Ni-doped $LaRhO_3$ is better

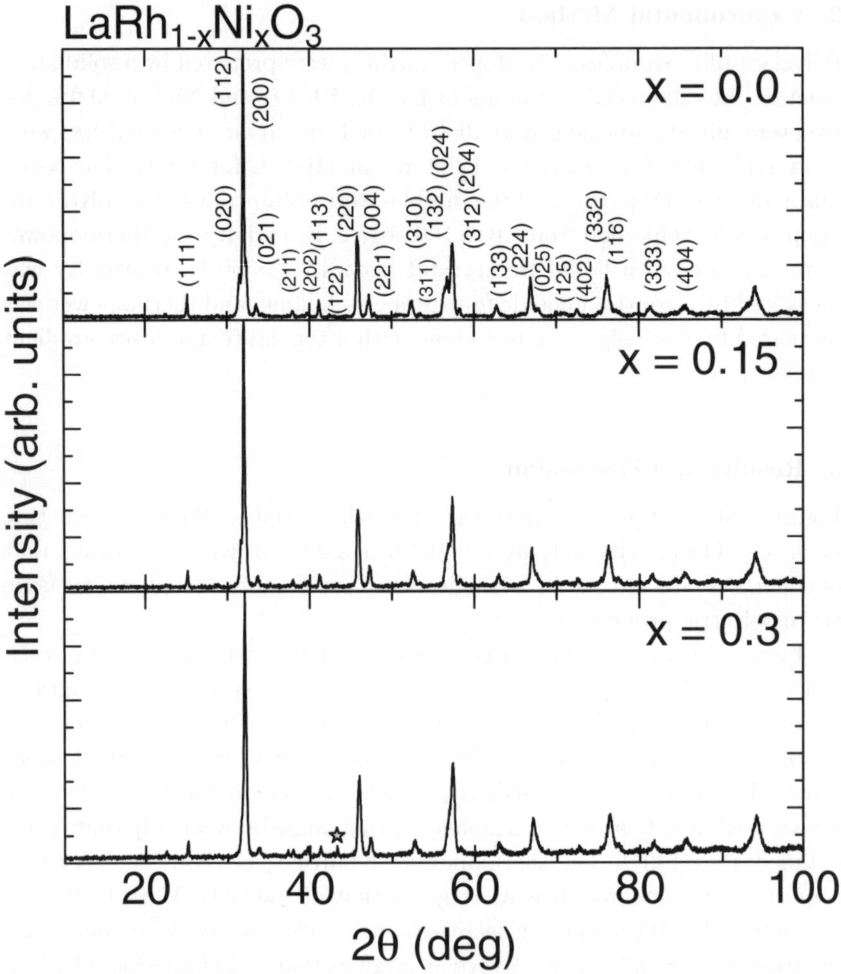

Fig. 2. XRD patterns of Ni-doped $LaRhO_3$. The peaks of $LaRhO_3$ are indexed as the $GdFeO_3$-type structure. A tiny peak of NiO impurity is observed only in $LaRh_{0.7}Ni_{0.3}O_3$.(marked with ⋆)

than Ni-doped $LaCoO_3$.

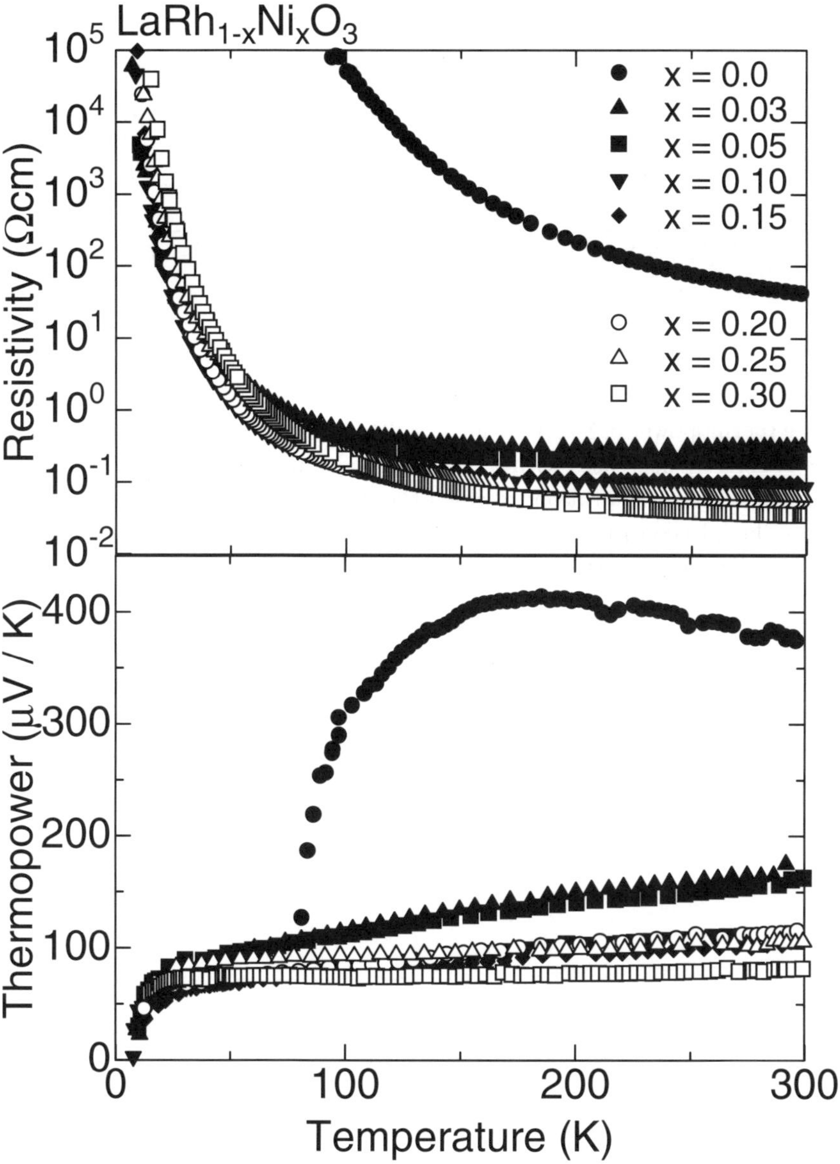

Fig. 3. Thermoelectric properties of Ni-doped $LaRhO_3$. Resistivity and thermopower below room temperature are shown.

4. Summary

In summary, we have measured the thermoelectric properties of Ni-doped $LaRhO_3$, and have compared them with those of Ni-doped $LaCoO_3$. The

Table 1. Comparison of $LaRh_{1-x}Ni_xO_3$[17] and $LaCo_{1-x}Ni_xO_3$[12,16,18].

		$LaRh_{1-x}Ni_xO_3$		$LaCo_{1-x}Ni_xO_3$	
		$x = 0.0$	$x = 0.3$	$x = 0.0$	$x = 0.3$
ρ(mΩcm)	300 K	4×10^4	36	2×10^4	6
	800 K	1×10^4	25	1.5	—
S(μV/K)	300 K	375	85	700	35
	800 K	400	185	50	—
κ(mW/cmK)	300 K	15	20	10	19

large thermopower of 85 μV/K for $x \geq 0.10$ is nearly independent of x while the resistivity decreases with x. In Ni-doped $LaRhO_3$, there may occur an electronic phase separation to which the doping-independent thermopower is ascribed. In contrast to Ni-doped $LaCoO_3$, Ni-doped $LaRhO_3$ shows good thermoelectric properties even at high temperature.

Acknowledgments

We thank T. Nakano and Y. Klein for fruitful discussion. We also thank S. Yoshida for technical support. This work was partially supported by a Grant-in-Aid for JSPS Fellows.

References

1. I. Terasaki, Y. Sasago and K. Uchinokura, *Phys. Rev. B* **56**, R12685 (1997).
2. R. Funahashi and I. Matsubara, *Appl. Phys. Lett.* **79**, 362 (2001).
3. Y. Miyazaki, K. Kudo, M. Akoshima, Y. Ono, Y. Koike and T. Kajitani, *Jpn. J. Appl. Phys.* **39**, L531 (2000).
4. A. C. Masset, C. Michel, A. Maignan, M. Hervieu, O. Toulemonde, F. Studer, B. Raveau and J. Hejtmanek, *Phys. Rev. B* **62**, 166 (2000).
5. W. Koshibae, K. Tsutsui and S. Maekawa, *Phys. Rev. B* **62**, 6869 (2000).
6. S. Okada and I. Terasaki, *Jpn. J. Appl. Phys.* **44**, 1834 (2005).
7. S. Okada, I. Terasaki, H. Okabe and M. Matoba, *J. Phys. Soc. Jpn.* **74**, 1525 (2005).
8. Y. Okamoto, M. Nohara, F. Sakai and H. Takagi, *J. Phys. Soc. Jpn.* **75**, 023704 (2006).
9. H. Kuriyama, M. Nohara, T. Sasagawa, K. Takubo, T. Mizokawa, K. Kimura and H. Takagi, *Proceedings of the 25th International Conference on Thermoelectrics (ICT2006).*
10. S. Shibasaki, W. Kobayashi and I. Terasaki, *Phys. Rev. B* **74**, 235110 (2006).
11. S. Shibasaki, W. Kobayashi and I. Terasaki, *Proceedings of the 26th International Conference on Thermoelectrics (ICT2007).*
12. R. Robert, L. Bochera, M. Trottmanna, A. Rellerb and A. Weidenkaffa *J. Solid Stat. Chem.* **179**, 3893 (2006).
13. Y. Tokura, Y. Okimoto, S. Yamaguchi, H. Taniguchi, T. Kimura and H. Takagi, *Phys. Rev. B* **58**, R1699 (1998).
14. R. D. Shannon, *Acta Cryst. B* **26**, 447 (1970).
15. T. Nakamura, T. Shimura, M. Itoh and Y. Takeda, *J. Solid State Chem.* **103**, 523 (1993).
16. P. Migiakis, J. Androulakis and J. Giapintzakis, *J. Appl. Phys.* **94**, 7616 (2003).
17. S. Shibasaki, Y. Takahashi and I. Terasaki, *submitted to Appl. Phys. Lett.*; *cond-mat* 0712.1626 (2007).
18. J. Hejtmánek, Z. Jirák, K. Knížek, M. Maryško, M. Veverka and C. Autret, *cond-mat* 0710.2762 (2007).

COLLECTIVE PRECESSION OF CHIRAL LIQUID CRYSTALS UNDER TRANSMEMBRANE MASS FLOW

GO WATANABE, SOU ISHIZAKI AND YUKA TABE

Graduate School of Science and Engineering, Waseda University
3-4-1 Okubo Shinjyuku-ku, Tokyo, 169-8555, Japan

Tilted monolayers composed of chiral liquid crystals (LCs) are known to exhibit the molecular collective precession under the transmembrane water flow. The phenomenon has been explained from the macroscopic point of view as a two-dimensional-type "Lehmann effect," but the microscopic mechanism has not been understood at all. In order to reveal the molecular motion, which could be the origin of the macroscopic precession, we calculated the rotational torque of an isolated chiral molecule under the mass flow of several materials. The result shows that single chiral molecule can obtain a certain torque under the mass flow along the molecular long axis, the magnitude of which depends on the chemical structure of flow materials.

1. Introduction

It would be an ultimate dream of human technology to create a perfectly controllable nano-scale machine. Stimulated by the recent researches on motor proteins [1], chemists started the trial to synthesize the molecular motors [2] and have successfully created the rotor molecules that can perform unidirectional rotation under some external fields such as chemical atmosphere [3] and photo-illumination [4]. The artificial molecular motors can transfer the non-rotating external fields to the unidirectional motion, but because their size is too small, they have not performed any practical tasks outside yet [2]. On the other hand, recent nanotechnology has dramatically miniaturized devices and machines, the size of which is now reaching sub-micrometer scale. Compared to such successful micromachines, nanomachines are much more influenced by thermal fluctuations, which should disturb them from working effectively.

It is likely that the real-functional artificial molecular motor can be synthesized in the near future, but we have tried to realize a nanomachine through another approach by using liquid crystalline molecular cooperation. Several years ago, we found that Langmuir monolayers composed of chiral liquid crystal compounds could exhibit the coherent unidirectional precession under transmembrane water transfer [5]. Since the thickness of the film is 2~3nm, if the area can be improved from the present 1cm^2 to less than 1 μm^2, the obtained molecular assembly will offer a new-type nanomachine.

In the liquid crystal area, the collective molecular rotation driven by external flow was first observed by O. Lehmann in 1900 [6]. He found that cholesteric LC droplets when heated from below exhibited violent rotation, resulting from the continuous molecular reorientation [6]. Named after the founder's name, such flow-induced molecular rotation has been called Lehmann effect, and the collective molecular precession in LC monolayers is regarded as its two-dimensional version. Theoretically, Lehman effect has been explained by phenomenological equations, in which the molecular rotation should be simply determined by macroscopic LC helicity and external flow direction[7]. While they could well reproduce the experimental results so far, there still remains an argument whether the microscopic interaction can be completely ignored. More recently, we found that the molecular rotation should depend on the molecular structure of flow materials as well [8], which suggests the essential role of the microscopic interaction in Lehmann effect. In order to investigate the microscopic origin of the collective molecular rotation, we calculated the torque of single chiral molecule submitted from the flow molecules by means of molecular dynamics (MD) simulation, and compared it with the experimental results.

2. Methods: MD Simulation

MD simulations were carried out by the MD program package Materials Explorer 4.0 (Fujitsu). We selected two LC compounds for the simulation; a chiral LC molecules, (2R)-2-[4-(5-octylpyrimidine-2-yl)phenyl-oxymethyl]-3-butyloxiran (hereafter, we call (R)-OPOB) shown in Fig. 1(a), and an achiral LC compound, 4'-hexyloxy-phenyl-2-(5-octyl)pyrimidine called P-6O8(Fig. 1(b)). The utilized molecular model was detailed atomic model. The potential function applying for the intramolecular interaction was Dreiding [9] and for the intermolecular interaction, we applied OPLS[10]. The total potential energy is written as

$$V_{total} = V_{stretch} + V_{angle} + V_{torsion} + V_{out-of-plane} + V_{vdW} + V_{Coulomb} \qquad (2)$$

The first term of the right hand equation indicates the bond-stretching potential;

$$V_{stretch} = \sum_{bonds} \frac{1}{2} k_r (r - r_0)^2 \qquad (3)$$

where k_r is the force constant, r_0 is the equilibrium bond length, and r is the actual bond length. The second term is the bond-angle potential;

$$V_{angle} = \sum_{angles} \frac{1}{2} k_\theta \left(\cos\theta - \cos\theta_0\right)^2 \tag{4}$$

where k_θ is the force constant, θ_0 is the equilibrium bond angle, and θ is the bond angle between three adjacent atoms. The third term gives the torsional potential;

$$V_{torsion} = \sum_{ijkl} \frac{1}{2} k_\phi \left\{1 - \cos\left[n\left(\phi_{ijkl} - \phi_0\right)\right]\right\} \tag{5}$$

where k_ϕ is the force constant, ϕ_0 is the equilibrium dihedral angle, and ϕ_{ijkl} is the dihedral angle formed by four consecutive atoms. The forth term is the out-of-plane potential;

$$V_{out-of-plane} = \sum_{ijkl} \frac{1}{2} k_\psi \left(\psi_{ijkl} - \psi_0\right)^2 \tag{6}$$

where k_ψ is the force constant, ψ_0 is the equilibrium out-of-plane angle, and ψ_{ijkl} is the out-of-plane angle formed by consecutive atoms. The second last term shows van der Waals interaction given by Lennard-Jones potential;

$$V_{vdW} = \sum_{i<j} \left(\frac{A_{ij}}{r_{ij}^{12}} - \frac{B_{ij}}{r_{ij}^{6}} \right) f_{ij} \tag{7}$$

where A_{ij} and B_{ij} are LJ coefficients and f_{ij} is taken as f_{ij} =0.5 for 1,4 terms, and f_{ij} =1 for all other terms. The last term is Coulomb potential written as

$$V_{Coulomb} = \sum_{i<j} \frac{1}{4\pi\varepsilon_0} \frac{q_i \cdot q_j}{r_{ij}} f_{ij} \tag{8}$$

where q_i is the electrical charge of the ith atom, r_{ij} is the distance between ith and jth atoms, and f_{ij} =0.125 for 1,4 terms, and f_{ij} =1 for all other terms. Before the MD simulation, we calculated atomic charges of each molecule by using MOPAC-PM5 (Fujistu) [11].

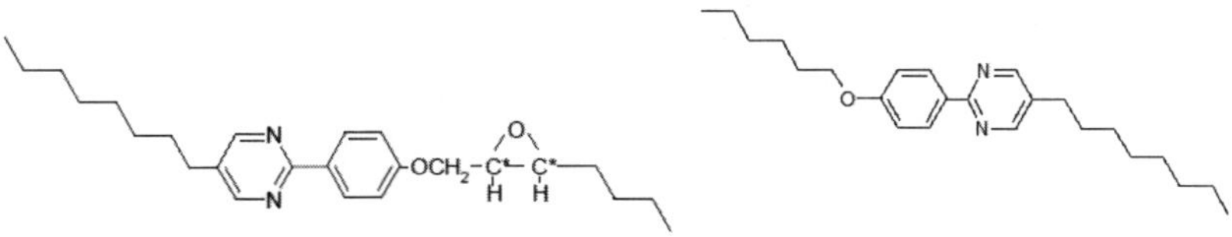

(a) (b)

Fig. 1. Structure of simulated LC molecules; (a) OPOB ((2R)-2-[4-(5-octylpyrimidine-2-yl)phenyl-oxymethyl]-3-butyloxiran), (b) P-6O8 (4'-hexyloxy-phenyl-2-(5-octyl)pyrimidine.

The system for each simulation is composed of one LC molecule and flow materials. The LC molecule was fixed as its long axis being parallel to z-axis in the middle of a rectangular MD cell (20.0 × 20.0 × 220.0 Å^3) under periodic boundary conditions. In the upper region of the cell (20.0 × 20.0 × 20.0 Å^3), flow molecules were placed as the initial state, which, under the constant force, moved toward –z direction colliding with the centered LC molecule. As the flow materials, we chose acetone, ethanol, ammonia and water, and the total collision numbers were 75 for acetone, 84 for ethanol, 192 for ammonia, and 280 for water. The initial velocities of flow materials were given by Maxwell-Boltzmann distribution of room temperature (298K) and they were pulled toward minus z-axis by the external force as mentioned above.

For the toque calculation, we calculated the rotational torque $\mathbf{N}(t)$ of the LC molecule around molecular long axis every step. When the i-th atom composed of the LC molecule is subjected to the force of $\mathbf{F}_i(t)$ from the flow, the torque around the molecular long axis is given by $\mathbf{r}_i(t) \times \mathbf{F}_i(t)$ where $\mathbf{r}_i(t)$ is the position vector of the i-th atom. The total torque is consequently given by

$$\mathbf{N}(t) = \sum_i \mathbf{r}_i(t) \times \mathbf{F}_i(t) \tag{1}$$

As the rotational torque around the molecular long axis, only the z-component of $\mathbf{N}(t)$ is taken for the analysis. In our simulation, we applied the strong force (~10^{-11}N) to the flow molecules so that all of them should completely transfer the LC molecule that allows us to compare the result of calculation to the experiment.

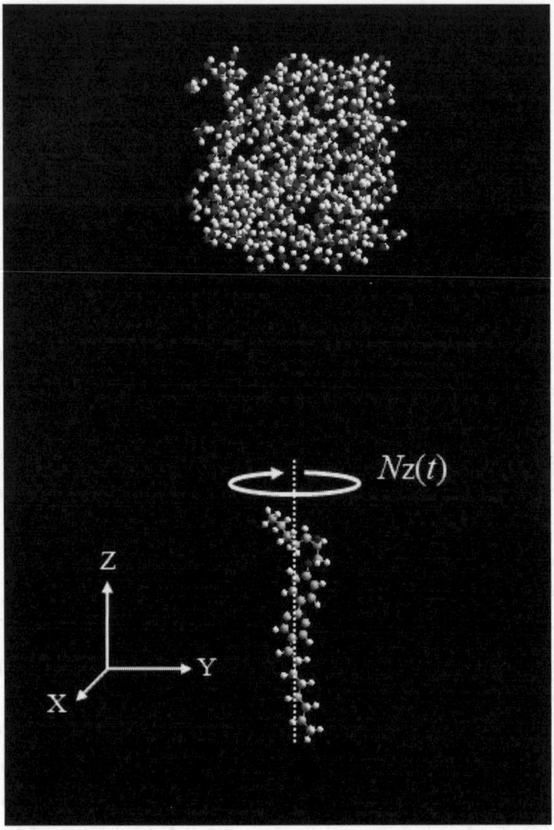

Fig. 2. Simulation model of MD consisting of a fixed LC molecule and water molecules.

3. Results and Discussion

3.1. *Torque Calculation for the Single Molecule*

Fig. 3(a) shows the torque given to the single OPOB molecule along the molecular long axis, caused by the water collision. The average value of (R)-molecule is $(-7.39 \pm 0.59) \times 10^{-14}$ J, and one of (S)-molecule is $(7.81 \pm 0.94) \times 10^{-14}$ J. Both (R)- and (S)- molecules obtain the non-zero torque in the average, the sign of which clearly depends on the chirality; (R)-isomer takes minus torque inducing CCW rotation from the flow direction, while the (S)-isomer is given plus torque causing the opposite rotation. Substituting the real experimental condition to the calculation, the torque value for the single molecule is $\sim 10^{-3}\ k_{\beta}$T.

This absolute value, being almost equal for each isomer, is much smaller than the thermal energy at room temperature, which indicates that the torque for the isolated molecule can be easily canceled unless it is magnified by the molecular cooperative correlation. Even though the rotational torque for the chiral OPOB molecules is small, it is qualitatively different from the case for achiral molecules. Fig. 3(b) shows the calculated torque for the achiral molecule P-6O8 with the same condition, where we find the torque is nearly zero. Since the molecular structure of OPOB is very similar to that of P-6O8 except for the chiral part, the obtained result suggests that the chirality should cause the molecular-level discriminative torque along the molecular long axis.

Next, we examined the influence of the flow materials. Fig. 4 shows the rotational torque submitted to the OPOB molecule by the collision of water, ammonia, ethanol and acetone. In terms of the sign of the torque, all the four flow materials show the same effect, while the efficiency is different. When the same number of flow molecules should collide the OPOB per unit of time, acetone can give the largest torque than the other three; four times larger than water or ammonia. The efficiency difference cannot be explained only by the molecular weight, but the chemical association should be taken into account.

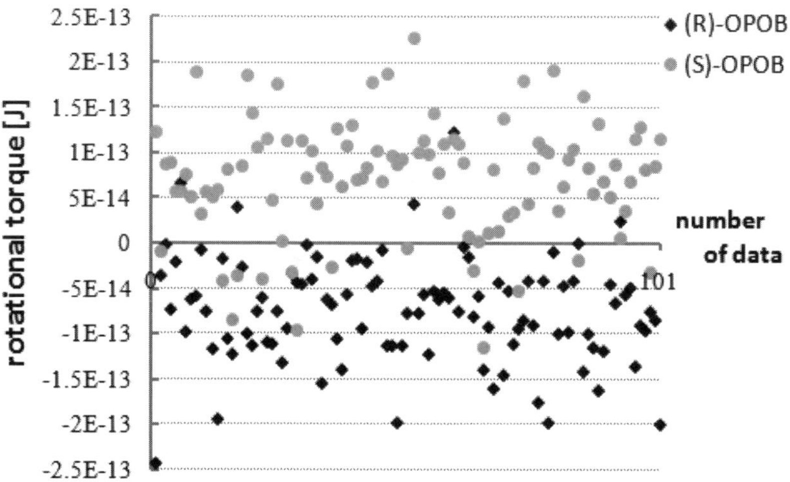

(a)

Fig. 3. Distribution graph of rotational torques caused by the water vapor of (a) chiral LC molecule OPOB and (b) achiral LC molecule P-6O8.

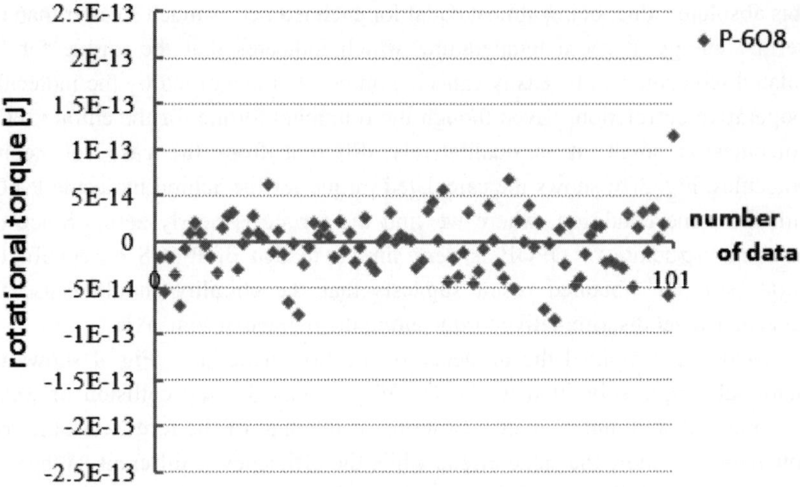

(b)

Fig. 3. (Cont'd)

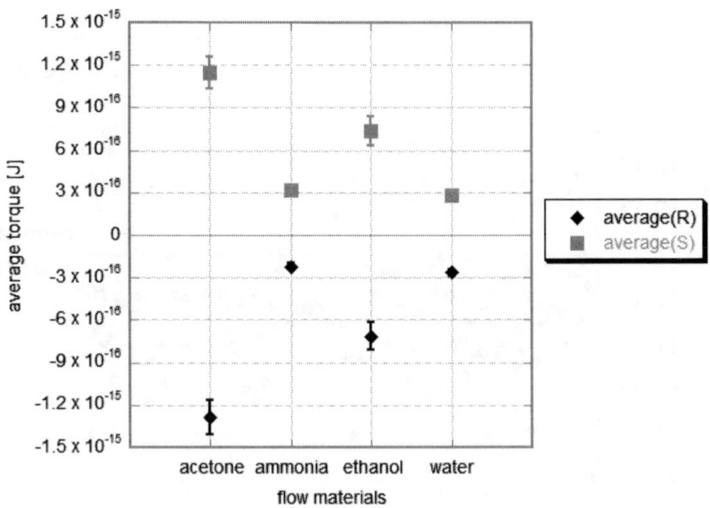

Fig. 4. Rotational torque of both R-isomer and S-isomer OPOB per unit molecule induced by each kind of flow materials.

3.2. *Discussion*

The torque calculated by the MD simulation qualitatively agrees with the experimental observation in LC monolayers[8]. When the OPOB monolayers are given the transmembrane flow of water, ammonia, ethanol and acetone , the unidirectional precession occurs in the same direction yet with the different speed, where acetone drives the fastest precession and the next is ethanol flowed by ammonia and water. Since the rotational torque calculated for the single LC molecule is much smaller than the thermal energy, it must be amplified by the liquid crystalline cooperation to cause the macroscopic precession. However, the qualitative agreement of the calculated torque and the macroscopic precession behavior suggests that the unidirectional collective precession should originate from each molecule's discriminative motion along the molecular long axis. From the symmetrical point of view, the chiral LC molecules should have a preference in the rotational direction when the up-down symmetry is broken by the flow.

As the next step, we need to reveal the mechanism how the molecular motion is transferred into the macroscopic precession motion. It is also important to calculate how many molecules are necessary to overcome the thermal fluctuation and transform the molecular motion into the macroscopic precession.

4. Summary

The two-dimensional Lehmann effect is investigated from the microscopic point of view by MD simulation. We calculated the rotational torque of single LC molecule, chiral OPOB or achiral P-6O8, induced by the flow of four different molecules of water, ammonia, ethanol and acetone. The result showed that OPOB molecule could obtain the discriminative rotational torque by the flow, while no considerable torque could be detected for P-6O8 even with the same condition. The sign of the torque depend on the molecular chirality but not on the flow materials, but the driving efficiency of the torque strongly depends on the molecular structure of the flow materials. The calculated torque for the single LC molecule well agrees with the experimental observation of the collective molecular precession in LC monolayers under the transmenbrane flow. The result suggests that the origin of the Lehmann rotation could be the discriminative motion of each LC molecule along the long axis.

Acknowledgments

We appreciate Drs. M. Yoneya and H. Yokoyama in LC nano-system project JST and Nanotechnology Research Institute AIST for their variable discussions.

References

1. M. Schliwa and G. Woehlke, *Nature* **422**, 759 (2003).
2. B. L. Feringa, *J. Org. Chem.* 72, 6635 (2007).
3. T.R. Kelly, H. De Silva, and R.A. Silva, *Nature* **401,** 150 (1999).
4. N. Koumura, R.W. J. Zijlstra, R.A. van Delden, N. Harada and B. L. Feringa, *Nature* **401,** 152 (1999).
5. Y. Tabe and H. Yokoyama, *Nature Materials* **2**, 806 (2003).
6. O. Lehmann, *Ann. Physik.*, **4**, 649 (1900).
7. F. M. Leslie, *Proc. Roy. Soc.*, **A307**, 359 (1968).
8. Y. Tabe, in preparation.
9. S. L. Mayo, B. D. Olafson, and W. A. Goddard III, *J. Phys. Chem.* **94**, 8897 (1990).
10. W. L. Jorgensen, J. D. Madura, and C. J. Swenson, *J. Am. Chem. Soc.* **106**, 6638 (1984).
11. J. J. P. Stewart, *MOPAC 2002 Manual Fujitsu Ltd* (2004).

INTERPLAY OF EXCITONS WITH FREE CARRIERS IN CARRIER TUNNELING DYNAMICS

SHULONG LU AND ATSUSHI TACKEUCHI

Department of Applied physics, Waseda University, Tokyo, 169-8555, Japan

SHUNICHI MUTO

Department of Applied Physics, Hokkaido University, Sapporo 060-8628, Japan

We present a systematic study of the interplay of excitons with free carriers in determination of carrier tunneling in GaAs/AlGaAs double quantum wells for various temperatures and carrier densities using time-resolved pump and probe measurements. The tunneling time abruptly decreases above a critical temperature (T_c) while it remains almost constant below T_c. As for the carrier density dependence, the tunneling time increases with increasing excitation power at low temperature. At high temperatures, no dependence of tunneling time on carrier density was observed. These behaviors are well explained in terms of the reduction of the electron tunneling probability of exciton states compared with that of free electron states, and the temperature and carrier density dependences of the populations of electrons in the two states. The thermodynamical equilibrium established between excitons and free carriers after photoexcitation determines the relative change of the population. Furthermore, the competition between exciton formation and carrier tunneling was discussed in a special case with 1 ps tunneling time.

1. Introduction

The excitonic nonlinearity in quantum wells (QWs) has attracted wide interest with regard to optical switching and ultrafast recovery.[1−2] Excitonic nonlinear effect arises from the fact that exciton formation rate is proportional to the product of free electron and hole densities.[3]In the past decades, the dynamics of excitons in QWs has been extensively studied.[4−8] Even in an asymmetric double QW (DQW) structure, exciton was also proved to play an important role in the tunneling process that carriers from one QW to an adjacent one through a thin potential barrier.[9−11]The interplay between excitons and free carriers in III-V semiconductors, and their relative contribution to the photoluminescence (PL) emission at the free-exciton energy, has recently been the subject of intense debate.[12–14] As a fundamental issue, the exciton formation time was reported to range from less than 10 ps up to about 1 ns.[3,6,8,15,16] The long formation time of excitons, together with the observation of PL at the exciton energy at the shortest time led Kira et al.[17,18] to introduce the idea that a free-electron-hole plasma, properly including Coulomb correlation effects, should give rise to PL at the exciton energy, without any exciton population. More investigation has indicated that both excitons and free carriers coexist and contribute to the PL. Also the temperature and carrier density are proved to be the two most critical parameters in determination of PL features.[8,12–14] On the basis of the same consideration, the competition between the exciton and free electron-hole pairs will have an avoidable effect on the tunneling process in the DQW. Due to the strong Coulomb interaction, carriers bound in excitonic states are more confined in the well. As a consequence, a slower tunneling rate of exciton is expected.[19]

In this paper, we systematically studied the tunneling process in the GaAs/AlGaAs DQW as a function of temperature and carrier density using time-resolved pump and probe reflection measurements. It was shown that the tunneling rate was influenced by the temperature and excitation power. The observed tunneling process can be explained by the interplay of excitons and free carriers determined by the thermodynamical equilibrium as a function of temperature and carrier density. Furthermore, the competition between exciton formation and carrier tunneling was discussed.

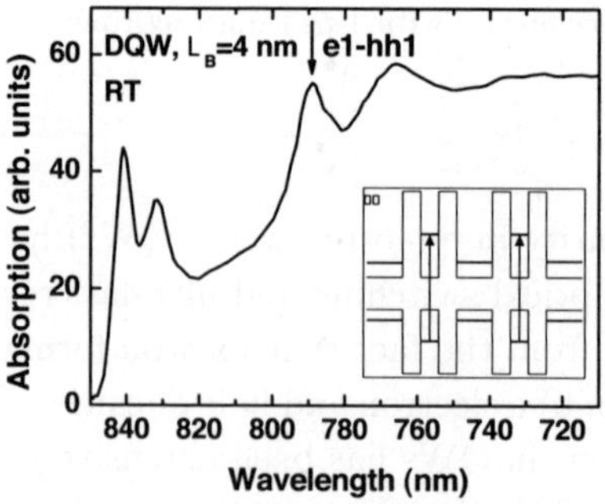

Fig. 1. Absorption spectrum for the DQW of $L_B = 4$ nm at RT. The arrow shows the lowest electron-heavy hole excitonic absorption peaks studied. Inset: Schematic energy band diagram.

2. Experimental Procedure

The samples under investigation consist of 50 periods of 4.5 nm (narrow) and 9.0 nm (wide) undoped GaAs QWs separated by $Al_{0.51}Ga_{0.49}As$ barriers. The thicknesses of the barriers, L_B, are 1.7, 2.8, 3.4 and 4 nm. The carrier tunneling occurs between two different energy levels in adjacent wells assisted by the LO phonon scattering.[20] For comparison, we also investigated a conventional MQW with a well width of 4.5 nm (the same as the narrow QW in the DQW) as well as a barrier width of 4 nm. Figure 1 shows the absorption spectrum of the DQW of L_B=4 nm. The schematic energy band diagram was shown in the inset. The absorption peak at about 840 nm shows the lowest electron-heavy-hole exciton peak of the wide well. The absorption peak at about 790 nm shows the lowest electron-heavy-hole exciton peak of the narrow well. The lowest conduction subband in the two adjacent wells is estimated to have an energy separation of about 60 meV. In the measurements, the samples were mounted on a cold finger cryostat with a temperature variation from 15 K to room temperature (RT) and a tunable optical pulse generated from a mode-locked Ti: sapphire laser with a pulse width of 100 fs and an 80 MHz repetition rate was used. The wavelengths of the pump and probe pulses were degenerate and tuned to satisfy the nearly resonant excitation of the absorption of the lowest electron-heavy-hole transition in the narrow QWs. The pump beam in front of the sample was focused onto a spot with a diameter of about 100 μm. The pump and probe experiment is usually considered to be caused by phase space filling and Coulomb interactions between optically excited electrons and holes. After the excitation of pump pulse, the time evolution of the carriers in the lowest electron-heavy hole excitonic state was probed by reflection measurement.

3. Experimental Results

Photoexcited carriers in the narrow QW of DQW can relax by tunneling into the wide well or recombine in the narrow well. Therefore, the measured decay rate ($1/\tau_{decay}$) in the DQW is determined by both tunneling rate ($1/\tau_t$) and recombination rate ($1/\tau_{rec}$) according to the equation of $1/\tau_{decay}=1/\tau_{rec}+1/\tau_t$. In this work, the tunneling time is defined as the inverse of the tunneling rate as described by above equation to avoid a controversial issue. Figure 2 shows the time-resolved reflection change in the narrow QWs of the four samples at RT. For each sample, the decay curve is composed of two parts. The fast relaxation on the short time range is

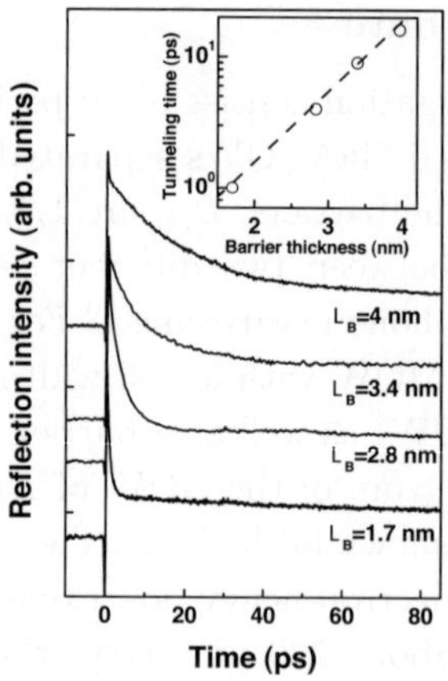

Fig. 2. Time-resolved reflection change in the narrow QWs of the four samples at RT. Inset: Electron tunneling times as a function of barrier thickness in a semi-logarithmic plot.

due to the electron tunneling from the narrow QWs to the wide ones. The tunneling time increases with increasing barrier thickness, as shown in the inset. The measured tunneling time through reflectance is identical to that obtained through transmission measurement[2] and is expected to depend exponentially on the barrier thickness.[21] Following the fast relaxation, a slow tail was observed. These tails have a magnitude of 30%-50% the initial reflection change. The pump-probe technique enables us to detect the change in hole population, therefore, the slow tail is attributed to the holes remaining in the narrow QWs. Furthermore, an investigation by S. Ten et al.[22] has indicated that the hole tunneling is crucially dependent on the in-plane momentum. A resonant excitation with zero excessive energy results in a slow hole tunneling.

Figure 3(a) shows the decay curves for the structure of L_B=4 nm at 15 K and RT. In the measurement, the excitation power is the same at every temperature. It can be found that the decay time due to the carrier tunneling is increased from 17 ps at RT to 42 ps at 15 K. In addition, the decay time of the slow part at 15 K is about 740 ps while 1.5 ns is observed at RT. The temperature dependence of the nonresonant tunneling times is shown in Fig. 3(b) with other two samples (2.8 and 3.4 nm). The tunneling time remains almost constant up to a critical temperature (T_c). Above the T_c, it falls markedly with increasing temperature. It is noted that the T_c is corresponding to the exciton binding energy.[23]The decay time at 15 K is about 2.5 times longer than that at RT.

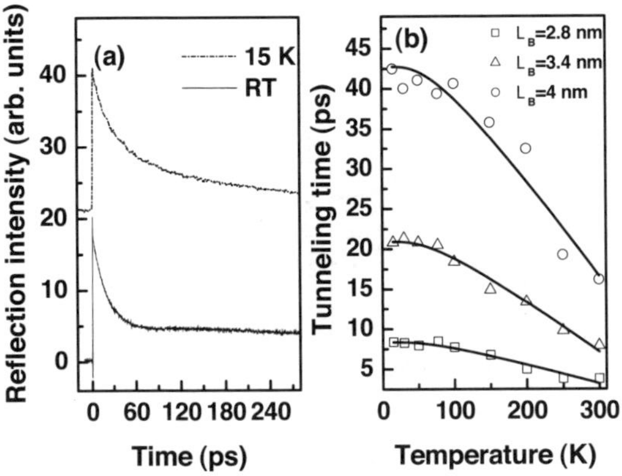

Fig. 3. (a) Time-resolved reflection intensity in the narrow QWs for the structure of $L_B = 4$ nm at 15 K and RT. (b) Temperature-dependent tunneling times for the structures with different barrier thicknesses. The solid lines show the calculated results.

Figure 4 shows the normalized decay curves for the DQW of L_B=4 nm as a function of excitation power at (a) 15 K and (b) RT. Before the normalization, we subtract the background signals for different excitation powers. At 15 K, the decay time due to electron tunneling increases with increasing excitation power. At RT, it becomes independent of the excitation power. For comparison, the recombination time of the MQW for different excitation powers was also measured at 15 K. Compared with the DQW, a much longer recombination time was observed and it decreased with increasing excitation power.[24] The decreased recombination time with increasing power in the MQW indicates that at a higher carrier density a larger fraction of the photogenerated carriers exists as excitons, resulting in an enhancement of the radiative recombination rate.[25] The carrier-density-dependent tunneling times (difference between the measured decay time in DQW and decay time in MQW) are plotted in Fig. 4(c), showing the tunneling times at 15 K and at RT. In the figure, the carrier density replaces the excitation power. Another sample of L_B=2.8 nm shows a similar way to this structure with increasing excitation power.

4. Discussion

The fast carrier tunneling between ground states in narrow and wide QWs is assisted by LO phonon scattering. However, LO phonon scattering has only a slight temperature dependence, $N_{ph}(T)+1$ $[N_{ph}(T)=1/(\exp(\hbar\omega_{LO}/k_BT)-1)]$, where N_{ph} (T) is the occupation of LO phonons as a function of tem-

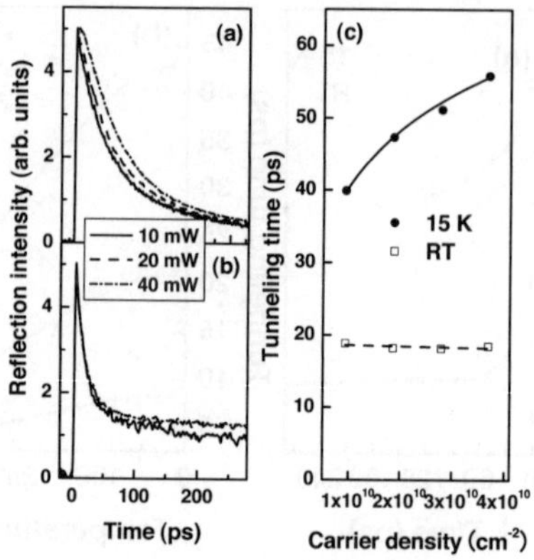

Fig. 4. Time-resolved change of reflection intensity in the narrow QW of the DQW of $L_B = 4$ nm at (a) 15 K and (b) RT for different excitation powers. (c) Tunneling time as a function of carrier density at 15 K and RT. The solid and dashed lines show the calculated results.

perature and $\hbar\omega_{LO}$ is the LO phonon energy in GaAs, too weak to account for the temperature dependence. As for the carrier-density-dependent tunneling behavior, the band filling and the electrostatic effects being due to spatial separation of electrons and holes can not explain the this behavior since the tunneling time is independent of the carrier density at RT. Taking into account of the contribution from the coexistence of excitons and free carriers to the tunneling process in the framework of rate equation, these phenomena can be well explained.

In a two-dimensional system, the thermodynamical equilibrium between the excitons and free carriers can be written, using the Saha equation,[8,15,26]

$$\frac{N_e N_h}{N_{ex}} = K = (\mu_x k_B T/\pi\hbar^2)\exp(-E_b/k_B T), \tag{1}$$

where μ_x is the exciton's reduced mass, k_B and $\hbar$ are Boltzmann's and Planck's constants, and E_b is the exciton binding energy. Electron and hole densities N_e and N_h, exciton density N_{ex} and total density of all carriers generated by a pump pulse per narrow well N, satisfy

$$N_e = N_h = N - N_{ex}. \tag{2}$$

Thus, the exciton population N_{ex} in terms of N is

$$\frac{N_{ex}}{N} = 1 + \frac{1}{2}\frac{K}{N} - \left(\frac{K}{N} + \frac{1}{4}\left(\frac{K}{N}\right)^2\right)^{1/2}. \tag{3}$$

From above equation, we can see that the temperature and the exciton binding energy determine the exciton population ratio.

Let us consider the time evolution of free carriers and excitons within the framework of rate equation analysis. After a δ-like excitation of pulse laser, the time evolution of free carriers and excitons can be written as[3]

$$\frac{dN_e}{dt} = -CN_eN_h - \frac{N_e}{\tau_e}, \tag{4}$$

$$\frac{dN_h}{dt} = -CN_eN_h - \frac{N_h}{\tau_h}, \tag{5}$$

$$\frac{dN_{ex}}{dt} = CN_eN_h - N_{ex}\left(\frac{1}{\tau_{rec}} + \frac{1}{\tau_{ex}}\right), \tag{6}$$

where CN_eN_h denotes the formation of an exciton and C is the formation coefficient; τ_{rec} is recombination time of excitons; τ_e, τ_h and τ_{ex} are the tunneling times of free electrons, free holes and electrons in excitonic states from the narrow QWs to the wide QWs, respectively. Here, we don't consider exciton ionization and free electron-hole recombination. For free carriers photoexcited in the narrow QW, the competition exists between the exciton formation and the tunneling. Following the hypothesis that exciton formation time is much faster than carrier tunneling time, from Eqs. (4) to (6), the time evolution of total carriers under steady-station follows the expression

$$\frac{N}{\tau} = \frac{N_e}{\tau_e} + \frac{N_{ex}}{\tau_{ex}} + \frac{N_{ex}}{\tau_{rec}} + \frac{N_h}{\tau_h}. \tag{7}$$

Since τ_e and τ_{ex} are much shorter than $\tau_{rec}$and τ_h, as a consequence, the decay curve includes nearly two components. The fast one is mainly determined by the tunneling time of free electrons and electrons in excitonic states, while the slow one is due to the recombination time of excitons and holes remaining. Because τ_{rec} and τ_h have a negligible modification to the tunneling time, we pay attention to the first two items in Eq. (7), i. e.,

$$\frac{N}{\tau_t} = \frac{N_e}{\tau_e} + \frac{N_{ex}}{\tau_{ex}}. \tag{8}$$

According to Eqs. (3) and (8), the temperature or carrier density-dependent tunneling time can be described as the following equation

$$\frac{1}{\tau_t} = \frac{1}{\tau_e} + \left(\frac{1}{\tau_{ex}} - \frac{1}{\tau_e}\right)\left(1 + \frac{1}{2}\frac{K}{N} - (\frac{K}{N} + \frac{1}{4}(\frac{K}{N})^2)^{1/2}\right). \tag{9}$$

Using Eqs. (1) and (9), the temperature or carrier density-dependent tunneling time is well fitted when a stable excitation power or temperature is given. Figures 3 and 4 presented the calculated results on the basis of the theoretical model. In the fitting, the tunneling times of excitons and free electrons are used as the fitting parameters. For the temperature-dependent tunneling time, the excitation power of 2×10^{10}cm^{-2} and exciton binding energy of about 10 meV are used. The fitted values of τ_e and τ_{ex} are 10 and 42 ps for the DQW of L_B=4 nm, 4 and 21 ps for the DQW of L_B=3.4 nm and 2.4 and 8.4 ps for the DQW of L_B=2.8 nm, respectively. The ratios of τ_e and τ_{ex} agree well with the previously predicted by Tackeuchi et al.[27] and are comparable to the previous results.[23] For the carrier density-dependent tunneling time, τ_e and τ_{ex} are 17.5 and 102 ps.[24] The ratio of τ_e to τ_{ex} is qualitatively agreement with the results obtained through the temperature-dependent tunneling time measurement. It is important to note that since we can't precisely know the real carrier temperature and carrier density, the fitted results can only support a rude qualitative understanding. At low temperature, excitons play an important role in the tunneling process, while free carriers are prevalent at high temperature due to the fast exciton thermalized ionization.[28]

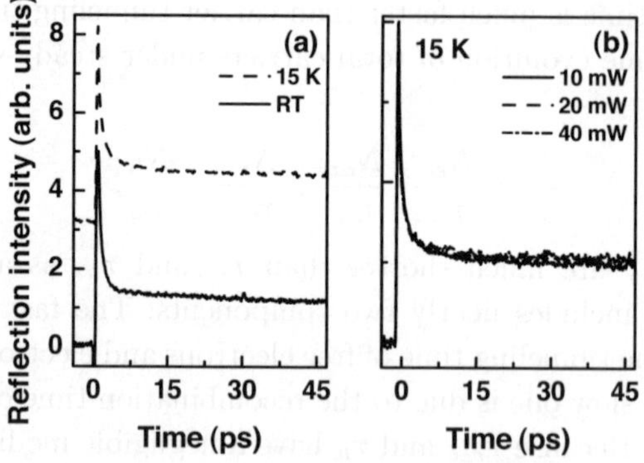

Fig. 5. Time-resolved reflection change in the narrow QWs for the structure of $L_B = 1.7$ nm for (a) different temperatures and (b) excitation powers at 15 K.

It is important to note that in the case of the sample of L_B=1.7 nm with 1 ps tunneling time at RT, we did not observe the same behavior as other three samples. Figure 5(a) shows the time-resolved reflection change in the narrow QWs at 15 K and RT. Besides the fast decay time due to carrier tunneling, the slow part does not show the change too. In addition, at 15 K, the decay curves remain unchanged with increasing excitation power, as shown in Fig. 5(b). As a fundamental issue, experimentally, the first investigation of exciton formation was studied by Damen et al.[6] Robart et al. defined that exciton formation time as the time to achieve the thermodynamical equilibrium between free carriers and excitons, starting from a system containing only free carriers. They reported a fast exciton formation time of less than 10 ps.[15] Considering the very fast tunneling time (1 ps), it is mostly likely that the exciton formation time is less than the carrier tunneling time of 1 ps. In this case, the time associated this tunneling process is shorter than the binding of the electron and the hole into an exciton, most of electrons of the narrow QWs will transfer to the wide QWs. Therefore, no temperature and carrier density-dependent behavior was observed.

5. Summary

We investigated the temperature and carrier density-dependent carrier tunneling process in GaAs/AlGaAs DQW. The prominent difference in tunneling time is due to the reduction of the electron tunneling probability of exciton states compared with that of free electron states, and the temperature and carrier density dependences of the populations of electrons in the two states. The thermodynamical equilibrium between excitons and free carriers determines the relative change of the population of the two kinds species. In addition, our experiment indicates that the exciton formation time is a critical point in determination the carrier tunneling process.

Acknowledgments

The authors would like to thank T. Ushiyama, T. Fujita, K. Kusunoki, M. Uesugi, and H. Nosho for their help in the measurement. This work was supported in part by a Grant-in-Aid for Scientific Research (No. 18360042), the 21st Century COE Program (Physics of Self-Organization Systems), Academic Frontier Project and the High-Tech Research Center Project from the Ministry of Education, Culture, Sports, Science and Technology of Japan.

References

1. H. M. Gibbs, S. S. Tarng, J. L. Jewell, D. A. Weinberger, K. Tai, A. C. Gossard, S. L. McCall, A. Passner, and W. Wiegmann, Appl. Phys. Lett. **41**, 221 (1982).
2. A. Tackeuchi, S. Muto, T. Inata, and T. Fujii: Appl. Phys. Lett. **58** 1670 (1991).

3. R. Strobel, R. Eccleston, J. Kuhl, and K. Kohler: Phys. Rev. B **43** 12564 (1991).
4. J. Feldman, G. Peter, E. O. Gobel, P. Dawson, K. Moore, C. Foxon, and R. J. Elliot, Phys. Rev. Lett. **59**, 2337 (1987).
5. H. Hillmer, A. Forchel, S. Hansman, M. Morohashi, E. Lopez, H. P. Meier, and K. Ploog, Phys. Rev. B **39**, 10901 (1990).
6. T. C. Damen, J. Shah, D. Y. Oberli, D. S. Chemla, J. E. Cunningham, and J. M. Kuo, Phys. Rev. B **42**, 12 (1990).
7. A. Vinattieri, J. Shah, T. C. Damen, D. S. Kim, L. N. Pfeiffer, M. Z. Maialle, and L. J. Sham: Phys. Rev. B **50** 10868 (1994).
8. H. W. Yoon, D. R. Wake, and J. P. Wolfe, Phys. Rev. B **54**, 2763 (1996).
9. Y. Sugiyama, A. Tackeuchi, T. Inata, and S. Muto, Jpn. J. Appl. Phys. **30**, L1454, (1991).
10. R. Ferreira, P. Rolland, Ph. Roussignol, C. Delalande, A. Vinattieri, L. Carraresi, M. Colocci, N. Roy, B. Sermage, J. F. Palmier, and B. Etienne, Phys. Rev. B **45**, 11 782 (1992).
11. I. Lawrence, S. Haacke, H. Mariette, W. W. Ruhle, H. Ulmer-Tuffigo, J. Cibert, and G. Feuillet, Phys. Rev. Lett. **73**, 2131 (1994).
12. S. Chatterjee. C. Ell, S. Mosor, G. Khitrova, H. M. Gibbs, W. Hoyer, M. Kira, S. W. Koch, J. P. Prineas, and H. Stolz, Phys. Rev. Lett. **92**, 067402 (2004).
13. J. Szczytko, L. Kappei, J. Berney, F. Morier-Genoud, M. T. Portella-Oberli, and B. Deveaud, Phys. Rev. B **71**, 195313 (2005).
14. A. Amo, M. D. Martin, L. Vina, A. I. Toropov and K. S. Zhuravlev, Phys. Rev. B **73**, 035205 (2006).
15. D. Robart, X. Marie, B. Baylac, T. Amand, M. Brousseau, G. Bacquet, G. Debart, R. Planel, and J. M. Gerard, Solid State Commun. **95**, 287 (1995).
16. R. A. Kaindl, M. A. Carnahan, D. Hgele, R. Lvenich, and D. S. Chemla, Nature (London) **423**, 734 (2003).
17. M. Kira, F. Jahnke, and S. W. Koch, Phys. Rev. Lett. **82**, 3544 (1999).
18. S. W. Koch, M. Kira, G. Khitrova, and H. M. Gibbs, Nature Material **5**, 523 2006.
19. V. Emiliani, S. Ceccherini, F. Bognai, M. Colocci, A. Frova, and S. S. Shi, Phys. Rev. B **56**, 4807 (1997).
20. S. Muto, T. Inata, A. Tackeuchi, Y. Sugiyama, and T. Fujii Appl. Phys. Lett. **58** 2393 (1991).
21. M. G. W. Alexander, M. Nido, W. W. Ruehle, and K. Koehler, Superlattice and Microstruc. **9**, 83 (1991).
22. S. Ten, M. F. Krol, P. T. Guerreiro, and N. Peyghambarian, Appl. Phys. Lett. **69,** 3387 (1996).
23. S. L. Lu, T. Ushiyama, T. Fujita, K. Kusunoki, A. Tackeuchi, and S. Muto, Jpn. J. Appl. Phys. **46**, 3305 (2007).
24. S. L. Lu, T. Ushiyama, A. Tackeuchi, and S. Muto, Phys. Stat. Sol. (c), **1-4**/DOI 10. 1002/pssc.200776579/(2007).
25. V. Srinivas, J. Hryniewicz, Y. J. Chen, and E. C. C. Wood, Phys. Rev. B **46**, 10193 (1992).

26. D. S. Chemla, D. A. B. Miller, P. W. Smith, A. C. Gossard and W. Wiegmann, IEEE J. Quantum Electron. QE **20**, 265 (1984).
27. A. Tackeuchi, S. Muto, T. Inata, and T. Fujii, Jpn. J. Appl. Phys. **28**, L1098 (1989).
28. W. H. Knox, R. L. Fork, M. C. Downer, D. A. B. Miller, D. S. Chemla, and C. V. Shank: Phys. Rev. Lett. **54** 1306 (1985).

PART C

Astrophysics as Interdisciplinary Science

NEW VIEW ON QUANTUM GRAVITY: MICRO-STRUCTURE OF SPACETIME AND ORIGIN OF THE UNIVERSE*

B. L. HU[†]

Department of Physics, University of Maryland, College Park, MD 20742 USA
[†] *E-mail: blhu@umd.edu*

It is generally agreed that the primary goal of quantum gravity is to find the microscopic structure of spacetime. However, for the last half a century the cardinal principle upheld by most general relativists has been to find ways to quantize Einstein's general theory of relativity, a theory which has proven to be highly successful in describing the macroscopic structure of spacetime we live in today. A tacit assumption in this existing paradigm is that doing so will yield the micro-structures of spacetime. We challenge this supposition and present a different view. If general relativity is an effective theory valid only at the long wavelength and low energy limits, and the metric and connection forms are collective variables, then quantizing a classical theory such as general relativity valid in the macroscopic domain is unlikely to yield a theory of the microscopic structures of spacetime. To uncover the microscopic structures one needs to find ways to unravel the underlying microscopic structures from observed macroscopic phenomena rather than naively quantizing the macroscopic variables, two very different paradigms. This task is similar to deducing the molecular constituents or even their quantum features from hydrodynamics or universalities of microscopic theories from critical phenomena. The macro to micro road poses a new and perhaps more difficult challenge to the next generation of theorists, phenomenologists and experimentalists in quantum gravity. Here we need to address issues at the quantum-classical and micro-macro interfaces familiar in mesoscopic physics, focusing on quantum fluctuations and correlations, coarse-graining and backreaction, and adopt ideas of nonequilibrium statistical mechanics and techniques from quantum field theory to explore theories built upon general relativity in a 'bottom-up' approach or a 'grass-root' road to quantum gravity. This view also provides us with a natural resolution towards the 'Origin of the Universe' issue, viz, the 'origin' is merely the commencement of a new phase where spacetime began to take shape and assume a manifold structure. This realization would push us to ask the more challenging question: How do we characterize the phase before? Was it a foam-like struc-

*This work is supported in part by National Science Foundation under grant PHY06-01550 ture of multiply-connected spactime? How did the phase transition take place – first or second order, discrete to continuum, stochastic to deterministic? An even more difficult question: How and 'when' did our concept of time and an arrow of time emerge whence we can ask how the universe came into being and evolved as far back as we can trace it?

1. Origin of the Universe and Quantum Gravity

1.1. *What is meant by the 'Origin' of the Universe?*

Based on observational evidence the general belief today is that our universe has been expanding, is adequately described by the Standard Model (based on the Robertson-Walker metric with a Friedmann solution) and had undergone a rapid exponential expansion much earlier, depicted by the inflationary model (Guth 1981). From Einstein's equations this implies that at a finite time in the past the universe must have been in a state of ultra-high curvature and mass density. This is what is generally referred to as the 'Big Bang' (BB). The theorists (Penrose, Hawking, Geroch 1967) who found it to be unavoidable in general relativity (GR) theory called this the cosmological 'singularity'.

Now, what exactly is meant by the 'origin' of the Universe? How did it come into being? Could it be avoided? And, more provocatively, what is *before the Big Bang*? These are the ultimate questions human beings are privileged enough to ask and to think. To address these questions requires some knowledge about the state, structure and dynamics of spacetime itself. We need a theory of the microscopic structures of spacetime.

The laws of classical gravity are believed to be valid from the scale of super-clusters of galaxies to the Planck scale, at 10^{-33} cm or 10^{-43} sec, an exceedingly small scale indeed. Smaller than this or earlier, we need to apply the laws of quantum physics, thus ushering in quantum gravity theory (QG), which for 30 years before modern string theory was born has been at one of the most challenging frontiers of theoretical physics. For most of that time, quantum gravity research focused on finding ways to quantize GR. The most developed theory from this vein is the loop quantum gravity program. The competing superstring theory has a very different character and most adherents believe that they have already found such a theory.

1.2. *What is Quantum Gravity? – Classical to Quantum, Macro to Micro*

The handful of schools may differ in their approaches to quantum gravity (see reading list), but they are likely to agree on one common agenda:

The goal of quantum gravity is to uncover the microscopic structure of spacetime. However, the way quantum gravity is defined or how the search is conducted varies a lot: Some believes that quantizing the macroscopic variables of spacetime (metric or connection) will yield the theory of microscopic structures. This has been pursued by general relativists for more than half a century. To others, spacetime is made up of strings or loops. Their tasks are to show how the familiar spacetime structure as we know it today arose. We can refer to this class of theories as the 'top down' models. Here I want to mention a third school of thought, which views the macroscopic variables of spacetime as *derived, collective* variables valid only at very low energies and large scales but loses meaning all together at much higher energies or smaller scales (e.g., smaller than the Planck length). This view advocates that one should forget about quantizing these macroscopic variables but search for the microscopic variables directly. Let me describe a view I favor.

1.3. *General Relativity as the Hydrodynamics of Spacetime Microstructure*

In this school of thought which begun with Sakharov in 1968, general relativity is viewed as the hydrodynamic (the low energy, long wavelength) regime of a more fundamental microscopic theory of spacetime, and the metric and the connection forms are the *collective variables* derived from them. At shorter wavelengths or higher energies, these collective variables will lose their meaning, much as the vibrational modes of a crystal will cease to exist at the atomic scale. If we view GR as hydrodynamics and the metric or connection forms as hydrodynamic variables, quantizing them will only give us a theory for the quantized modes of collective excitations, such as phonons in a crystal, but not a theory of atoms or quantum electrodynamics (QED) which is a more fundamental theory.

According to this view most macroscopic gravitational phenomena can be explained as collective modes and hydrodynamic excitations, from gravitational waves as weak perturbations, to black holes in the strong regime, as solitons. With better observational tools or numerical techniques available, we may even find analogs of turbulence effects in this geometrohydrodynamics.

This view is very different in meaning and in practice from the other approaches. To better appreciate the roots of these differences, let us step back for a moment and review the two major paradigms in physics which underscore the two main directions of research in theoretical cosmology.

2. Two Major Paradigms in Physics and Two Directions of Research in Cosmology

There are two basic aspects in the formulation of any cosmological model. One aspect involves the basic constituents and forces, the other involves the structure and dynamics, i.e., the organization and processing of these constituents as mediated by the basic forces or their derivatives. The first aspect is provided by the basic theories describing spacetime and matter. The second aspect in addressing the universe and its constituents is cosmology proper.

2.1. *Two paradigms: Elementary Particle versus Condensed Matter Physics*

It is not difficult to recognize that actually these two aspects permeate throughout almost all subfields in physics, or science in general. Examples of the first aspect in physics dealing with the "basic" constituents and forces are general relativity, quantum field theory, quantum electrodynamics (QED), quantum chromodynamics (QCD), grand unified theories, supersymmetry, supergravity and superstring theories. The second aspect dealing with the structure and dynamics is the subject matter of biology, chemistry, molecular physics, atomic physics, nuclear physics and particle physics. The former aspect is treated today primarily in the disciplines of elementary particle physics and quantum gravity. The latter aspect is treated today in the discipline known collectively as condensed matter physics. In this sense we can, for example, regard nuclear physics as the condensed matter physics of quarks and gluons.

Note, however, the *duality* and the *interplay* of these two aspects in any discipline. On the one hand the basic laws of nature are often discovered or induced from close examination of the structure and properties of particular systems - witness the role played by atomic spectroscopy and scattering in the discovery of quantum mechanics and atomic theory, accelerator experiments in advancing particle physics. On the other hand, once the nature of the fundamental forces and constituents are known, one attempts to depict reality by deducing possible structures and dynamics from these basic laws. Thus the study of electrons and atoms via electromagnetic interaction has been the underlying fabric of condensed matter physics for a while. Deducing nuclear force from QCD remains the central task of nuclear research today. From general relativity one attempts to deduce the properties of neutron stars, black holes and the universe, which is the theme of relativistic astrophysics and cosmology.

Note that many known physical forces are not *fundamental* (in the sense that they are irreducible), but are *effective* in nature. Molecular forces and nuclear forces are such examples. One may also regard gravity as an effective force. Note also that many disciplines contain dual aspects. This is especially true in the developing areas, in which the basic forces and constituents of the system are not fully understood. For example, particle physics deals both with the structure and the interactions as quantum flavor and color dynamics. The duality of compositeness and elementarity should also be present in superstring and other theories of quantum gravity.

2.2. *Two Aspects in Cosmology*

What about cosmology? The above-mentioned dual aspects are certainly apparent. What is new is that in addition to matter (as described by particles and fields) we have to add in the consideration of spacetime (as described by geometry and topology). In the first aspect concerning constituents and forces, there are also two contrasting views. The "idealist" takes the view that spacetime is the basic entity, the laws of the universe is governed by the dynamics of geometry. Matter is viewed as perturbations of spacetime, particle as geometrodynamic excitons. These ideas, as extension of Einstein's theory, are not that strange as they may appear: Particles are representations of internal symmetries, graviton the resonant modes of strings. By contrast, the "materalist" takes the view that spacetime is the manifestation of collective, large scale interaction of matter fields. Thus according to Sakharov, gravity should be treated as an effective theory, like elasticity to atomic forces. This is expressed in the induced gravity program. Despite its many technical difficulties, this view still evokes some sobering thoughts. It suggests among others that the attempt to deduce a quantum theory of gravity by quantizing the metric may prove to be as meaningful as deducing QED from quantizing elasticity.

In recent years the apparent contrast between particle-fields and geometry-topology has dissolved somewhat in the wake of superstring theory. The fact that the same concept can be viewed in both ways and that spacetime and strings appear in different regimes may indeed offer some new insights into the fundamental aspects of our universe. Duality between the high and low energy sectors, correspondence between gauge theory in the bulk and conformal field theory on the boundary and the holography principle of meaningful information residing on the surface are perhaps some of the most attractive ideas evolved.

As for the second aspect in cosmology, i.e., the manifestation of basic forces in astrophysical and cosmological processes, one sees that almost with any subdiscipline of physics there is a corresponding branch of astrophysics. However, the central theme of cosmology which addresses the state of the universe as a whole is more than the sum-total of its individual components, as depicted by the many subdisciplines of astrophysics. There are broader issues special to the overall problem of how the universe comes into being and why it should be the way it is, which may be traced back to the puzzling issues at the foundation of quantum mechanics and general relativity. The inquiry of these issues will necessarily bring us back to the 'fundamental' aspect of inquiries discussed above.

2.3. *Two Directions of Cosmological Research*

Depending on the relative emphasis one puts on these two aspects, current research on cosmological theories follow roughly two directions:
A) Cosmology as consequences of quantum gravity and superstring theories.
B) Cosmology describing the structure and dynamics of the universe

In this first direction one could also include inquiries or proposals made which view the universe as manifestor of physical laws, as formulator of rules, as processor of information, etc. This direction of cosmological research touches on the basic laws of quantum mechanics, general relativity and statistical mechanics. In this field the formulation of meaningful problems are almost as important as seeking their solutions. Progress will be slow but the intellectual reward is profound.

The second direction is characterized in my figurative depiction "Cosmology as 'Condensed Matter' Physics", the title of a paper for a conference held in Hong Kong in 1987. By "condensed state" I refer both to matter and spacetime. Cosmology is the study of the organization and processing of matter as well as spacetime points. I presented via several tables in that article some major ingredients of condensed matter physics, nuclear physics and the physics of the early universe and outlined the major themes of recent development of condensed matter physics. Notice the increasing importance attached to nonlinear, nonlocal and stochastic behavior of complex systems. In my opinion, two new ingredients will likely play a dominant role in the structure and evolution of the primordial universe at the Planck scale: One is topology and the other is stochasticity, both for matter-field and spacetime-geometry.

New impetus is provided by advances both in 1) particle physics and quantum gravity, such as superstring theory, loop quantum gravity and simplicial gravity, where close mathematical formulation of these problems has become possible, and 2) condensed matter physics such as critical dynamics and quantum phase transition, order-disorder cross-over, dynamical and complex systems, etc. Developing the synergy between these two major disciplines or sectors of physics can open up new possibilities in probing the organization and dynamics of matter in various states. These techniques and ideas may also provide useful hints in understanding how spacetime takes shape, how the universe evolves, what determines its content and how its many different structural forms develop. Cosmological research would benefit from recognizing and harnessing these resources.

3. The Mescoscopic Structures of Spacetime and Stochastic Gravity

As remarked in Sec. 2, our view of quantum gravity is that we find it more useful to find the micro-variables than to quantize macroscopic variables. And of the two paradigms described above, in order to unravel the microscopic structure of spacetime, our outlook is closer to the condensed matter than the elementary particle physics paradigm. Our approach relies more on statistical and stochastic methods and our focus will be on how the known macroscopic levels of structure emerge from the unknown underlying substructures. If we view classical gravity as an effective theory, i.e., the metric or connection functions as collective variables of some fundamental constituents which make up spacetime in the large, and general relativity as the hydrodynamic limit, we can also ask if there is a regime like kinetic theory of molecular dynamics or mesoscopic physics of quantum many body systems intermediate between quantum micro-dynamics and classical macro-dynamics. In addition to serving many practical applications, mesoscopic physics also embodies some fundamental issues. It dwells on two central issues in theoretical physics: the micro to macro and the quantum to classical transitions.

To identify any intermediate levels of structure of spacetime between the macro and the micro, it is useful to examine the existing gravitational theories beginning with GR.

3.1. *The three lowest layers: Classical, Semiclassical and Stochastic Gravity*

The theory of general relativity provides an excellent description of the features of large scale spacetime and its dynamics. *Classical gravity* assumes classical matter as source in the Einstein equation. When quantum fields are included in the matter source, a quantum field theory in curved spacetimes is needed. At the semiclassical level the source in the Einstein equation is given by the expectation value of the energy momentum tensor operator of quantum matter fields with respect to some quantum state. *Semiclassical gravity* refers to the theory where classical spacetime is driven by quantum fields as sources, thus it includes the backreaction of quantum fields on spacetime and the evolution of quantum field and spacetime self-consistently. This is the theory where fundamental discoveries in black hole physics by example of Hawking radiation and in cosmology such as the inflationary universe were made. This serves as a solid platform for us to start the expedition towards quantum gravity, which emphasized, is a theory of the microscopic structure of spacetime, not necessarily obtainable from quantizing general relativity. The next higher layer is *stochastic gravity* which includes the fluctuations of quantum field as source described by the Einstein-Langevin equation. Our approach to *quantum gravity* uses stochastic gravity as a launching platform and mesoscopic physics as a guide.

So what is the physics of mesoscopic spacetime and how does stochastic gravity enter?

3.2. *Mesoscopic Structure and Stochastic Gravity*

In a 1994 conference paper I pointed out that many issues special to this intermediate stage between the macro and micro structures, such as the transition from quantum to classical spacetime via the decoherence of the 'density matrix of the universe', phase transition or cross-over behavior at the Planck scale, tunneling and particle creation, or growth of density contrast from vacuum fluctuations, share some basic concerns of mesoscopic physics in atomic/optical, particle/nuclear and condensed matter or quantum many body systems. Underlying these issues are three main factors: quantum *coherence, fluctuations and correlations.* We discuss how a deeper understanding of these aspects of fields and spacetimes related to the quantum / classical and the micro / macro interfaces, the discrete / continuum or the stochastic / deterministic transitions can help to address some basic problems in gravity, cosmology and black hole.

Stochastic gravity is a consistent and natural generalization of semiclassical gravity to include the effects of quantum fluctuations. The centerpiece of this theory is the stress-energy bi-tensor and its expectation value known as the noise kernel. We believe that precious new information resides in the two-point functions and higher order correlation functions of the stress energy tensor which, through the generalized Einstein-Langevin equations for the higher-order induced metric correlations, may reflect the finer structures of spacetime at a scale when information provided by its mean value as source (semiclassical gravity) is no longer adequate. The key point here is the important role played by noise, fluctuations, dissipation, correlations and quantum coherence, the central issues focused on by mesoscopic physics. Noise carries information about the correlations of the quantum field and the quantum coherence in the gravity sector is obtained from the correlations of induced metric fluctuations. Stochastic gravity provides a relation between noise in quantum fields and metric fluctuations.

This new framework allows one to explore the quantum statistical properties of spacetime: How fluctuations in the quantum fields induce metric fluctuations and seed the cosmic structures, quantum phase transition in the early universe, black hole quantum horizon fluctuations, stochastic processes in the black hole environment, the back-reaction of Hawking radiance in black hole dynamics, and implications on trans-Planckian physics. On the theoretical issues, stochastic gravity is the necessary foundation to investigate the validity of semiclassical gravity and the viability of inflationary cosmology based on the appearance and sustenance of a vacuum energy-dominated phase. It is also a useful platform supported by well-established low energy (sub-Planckian) physics to explore the connection with high energy (Planckian) physics in the realm of quantum gravity.

3.3. *Spacetime as an Emergent Collective State of Strongly Correlated Systems*

Viewing the issues of correlations and quantum coherence in the light of mesoscopic physics we see that what appears as a source in the Einstein-Langevin equation, the stress-energy two point function, is analogous to conductance of electron transport which is given by the current-current two point function. What this means is that we are really calculating the transport functions of the matter particles as depicted here by the quantum fields. Following Einstein's observation that spacetime dynamics is determined by (while also dictates) the matter (energy density), we expect that the transport function represented by the current correlation in the fluctuations of the matter energy density would also have a geometric counterpart and equal significance at a higher energy than the semiclassical gravity scale. This is consistent with general relativity as hydrodynamics: conductivity, viscosity and other transport functions are hydrodynamic quantities. Here we are after the transport functions associated with the dynamics of spacetime structures. The Einstein tensor correlation function in Minkowsky spacetimes calculated by Martin and Verdaguer is a first step. Another example is given by Shiokawa who computed the metric conductance fluctuations.

For many practical purposes we don't need to know the details of the fundamental constituents or their interactions to establish an adequate depiction of the low or medium energy physics, but can model them with semi-phenomenological concepts. When the interaction among the constituents gets stronger, or the probing scale gets shorter, effects associated with the higher correlation functions of the system begin to show up. Studies in strongly correlated systems give revealing examples. Thus, viewed in the light of mesoscopic physics, aided by stochastic gravity, we can begin to probe into the higher correlations of quantum matter and with them the associated excitations of the collective modes in geometro-hydrodynamics, the kinetic theory for the meso-dynamics of spacetime, and eventually quantum gravity – the theory of micro-spacetime dynamics.

In seeking a clue to the micro theory of spacetime from macroscopic constructs, we have focused here on the kinetic / hydrodynamic theory and noise / fluctuations aspects. Statistical mechanical and stochastic / probability theory ideas will play a central role. We will encounter nonlinear and nonlocal structures (nonlocality in space, nonMarkovian in time) in abundance. Another equally important factor is topology: Topological features can have a better chance to survive the coarse-graining or effective / emergent processes to the macro world and can be a powerful key to unravel the hidden structures of the microscopic world.

4. One Vein of the 'Hydro' View: Spacetime as Condensate?

Viewing our spacetime as a hydrodynamic entity, I'd like to explore with you a new idea inspired by the development of Bose-Einstein Condensate (BEC) physics in recent years. The idea is that, maybe spacetime, describable by a differentiable manifold structure; valid only at the low-energy long-wavelength limit of some fundamental theory, is a condensate. We have examined what a condensate means in a recent essay, but for now we can use the BEC analog and think of it as a collective quantum state of many atoms with macroscopic quantum coherence.

4.1. *Unconventional view 1: All sub-Planckian physics are low temperature physics*

Atom condensates exist at very low temperatures. It takes novel ways of cooling the atoms, many decades after the theoretical predictions, to see a BEC in the laboratories. It may not be too outlandish to draw the parallel with spacetime as we see it today, because the present universe is rather cold (~3K). But we believe that the physical laws governing today's universe are valid all the way back to the GUT (grand unification theory) and the Planck epochs, when the temperatures were not so low any more. Any normal person would consider the Planck temperature $T^{Pl} = 10^{32}$K a bit high. Since the spacetime structure is supposed to hold (Einstein's theory) for all eras below the Planck temperature,

if we consider spacetime as a condensate today, shouldn't it remain a condensate at this ridiculously high temperature?

YES is my answer to this question. What human observers consider as high temperature (such as that when species homo-sapiens will instantly evaporate) has no effect on the temperature scales defined by physical processes which in turn are governed by physical laws. Instead of conceding to a breakdown of the spacetime condensate at these temperatures, one should push this concept to its limit and not be surprised at the conclusion that all known physics today, as long as a smooth manifold structure remains valid for spacetime, the arena where all physical processes take place, are low- temperature physics. Spacetime condensate began to take shape at the Planck temperature, but will cease to exist above it, according to our current understanding of the physical laws. In this sense spacetime physics as we know it is low temperature hydrodynamics, and, in particular, today we are dealing with ultra-low temperature physics, similar to superfluids and BECs.

4.2. *Unconventional view 2: Spacetime is, after all, a quantum entity*

An even more severe difficulty in viewing spacetime as a condensate is to recognize and identify the quantum features in spacetime as it exists today, not at the Planck time. The conventional view holds that spacetime is classical at scales larger than the Planck length, but quantum if smaller. That was the rationale for seeking a quantum version of general relativity, beginning with quantizing the metric function and the connection forms. The spacetime condensate view holds that the universe is fundamentally a quantum phenomenon, but at the mean field level the many body wave functions (of the micro-constituents, or the 'atoms' of spacetime) which we use to describe its large scale behavior (order parameter field) obey a classical-like equation, similar to the Gross-Pitaevsky equation in BEC, which has proven to be surprisingly successful in capturing the large scale collective dynamics of BEC, until quantum fluctuations and strong correlation effects enter into the picture.

Could it be that the Einstein equation depicting the collective behavior of the spacetime quantum fluid is on the same footing as the Gross-Pitaevsky equation for BEC? The deeper layer of structure is ostensibly quantum; it is only at the mean field level that the many-body wave function is amenable to a classical description. We have seen many examples in quantum mechanics where this holds. For any quantum system which has bilinear coupling with its environment or is itself Gaussian exact (or if one is satisfied with a Gaussian approximation description) the equations of motion for the expectation values of the quantum observables have the same form as its classical counterpart. Ehrenfest theorem interconnecting the quantum and the classical is an example.

The obvious challenge is, if the universe is intrinsically quantum and coherent, where can one expect to see the quantum coherence phenomena of spacetime? Here again we look to analogs in BEC dynamics for inspiration, and there are a few useful ones, such as particle production in the collapse of a BEC (Bosenova) experiment. One obvious phenomenon staring at our face is the vacuum energy of the spacetime condensate, because if spacetime is a quantum entity, vacuum energy density exists unabated for our present day late universe, whereas its origin is somewhat mysterious for a classical spacetime in the conventional view.

5. Implications for the Origin of the Universe and other Issues

So what does this alternative viewpoint of quantum gravity say about the important issues of cosmology? Let's begin with the low energy phenomena accessible by our observations today. There are many discussions these days about Lorentz invariance being broken at ultra-high energies, from superstring and other theories. Lorentz symmetry is a well established symmetry of local (Minkowsky) structure of our spacetime, first found in Maxwell's equations of electromagnetism, later adopted as the symmetry of special relativity as the new laws of mechanics which displaces the Galilean symmetry underlying Newton's theory. Minkowsky provided the geometric description of this new theory of spacetime.

In the 'hydro' spacetime viewpoint, Lorentz invariance is an emergent symmetry which applies only to our large scale structure of spacetime at this late stage of its evolution, not unlike the many symmetries of fluids which are not apparent at the molecular dynamics level, which, at a finer scale, its structure and dynamics are governed by a different set of symmetries. Lorentz and other symmetries came into being at sub-Planckian energy, when spacetime took on a form continuous and smooth enough so that a manifold structure appears and differential geometric description became applicable. At sub-Planckian length it could be in a foam-like structure with non-trivial topologies, as captured in the spacetime foam idea of Wheeler. Lorentz as well as many symmetries associated with a smooth large scale manifold structure will be completely out of place. So it is not such a startling thing to imagine giving up many of our much cherished law and order, because our experience is limited to very special conditions. From the emergent viewpoint or from the philosophy of an 'ephemeralist' (e.g., Zuang Zhou) no laws are sacrosanct and nothing is for ever.

Another implication of this view is that with a clear vision that there exists fundamental differences in the levels of structure, it can help us identify untenable ideas and not to waste time in meaningless pursuits. Before describing activities involving spacetime and its more elementary constituents, one should explain where/how spacetime structure comes into being. For example, 'string cosmology' as a research area appears strange to me: One cannot just write down a metric, throw strings into it and start proposing a cosmological theory that way. How could strings propagate on a structure as yet to be determined by their interactions? One needs to show how strings made up our spacetime, or at the least, identify the particular regime(s) where strings are free to propagate. In fact, despite the often mention of inflation in many papers on string cosmology, the major players of string theory who are honest in reporting their results are likely to agree that they have not been successful in predicting inflation from string theory. Similarly, speculations into pre-Big Bang, if BB signifies the beginning of spacetime, would be useless if it invokes a manifold structure for the background spacetime.

On the constructive aspects, this view could provide a different and perhaps better approach to important issues, such as the mystery of dark energy: Why is the cosmological constant so low (compared to natural particle physics energy scale) today, and so close to the matter energy density (the coincidence issue)? We have also explored the implications of this view on quantum mechanics and general relativity, and its relation to string theory and loop quantum gravity. You can read more about this in my essay on spacetime condensates and Professor Volovik's book.

Finally, what would this view say about the origin of the Universe? I think phase transition is probably a better way to address this question. We live in a low energy (temperature) phase and our description of spacetime is good only at its very long wavelength limit. The phase transition point according to our present understanding of physics could be the Planck energy (remember we said that even slightly below the Planck temperature is considered low temperature). In this view the origin of the universe is really just the beginning of a new low energy phase, not unlike water turning into ice below the freezing point. For creatures incapable of living in a subzero temperature they would regard the freezing point as the beginning of their universe.

Now, what came before that? In this view it is not difficult to imagine there is an epoch before this 'beginning', in fact, perhaps many different epochs and many beginnings – no mysticism here. Before the Big Bang, or if you like sensational advertisements, the 'Birth of the (our) Universe', there existed a different phase in the structure of spacetime. We need a different set of variables to describe its basic constituents (not metric or connection), a different language to capture its structure (not differential geometry) and a different set of equations to describe its dynamics (not general relativity). Unlike the summer insects, even if we cannot live that phase we can surely think about its attributes and even devise tools to capture its essence. That is the power of human intellect, will and spirit imbued in the fearless quests of theoretical physics, that which brought you and me together here today.

References

The key ideas in this article are expressed in these essays:

1. "GENERAL RELATIVITY AS GEOMETRO-HYDRODYNAMICS" Invited talk at the Second Sakharov International Conference, Lebedev Physical Institute, May, 1996. [gr-qc/9607070].
2. "SEMICLASSICAL GRAVITY AND MESOSCOPIC PHYSICS" Invited talk at the International Symposium on Quantum Classical Correspondence, Drexel University, Philadelphia, Sept. 1994. Proceedings eds D. H. Feng and B. L. Hu (International Publishers, Boston, 1997) [gr-qc/9511077]
3. "COSMOLOGY AS 'CONDENSED MATTER' PHYSICS" Invited talk given at the Third Asia-Pacific Physics Conference, Hong Kong,

June 1988. Proceedings edited by K. Young (World Scientific Publishing Co., Singapore, 1989). [gr-qc/9511076]
4. "A KINETIC THEORY APPROACH TO QUANTUM GRAVITY" Invited talk given at the 6th Peyresq Meeting, France, June 2001. Int. J. Theor. Phys. **41** (2002) 2111-2138 [gr-qc/0204069]
5. "CAN SPACETIME BE A CONDENSATE?" Int. J. Theor. Phys. 44 (2005) 1785 [gr-qc/0503067]

Readings:
Approaches to Quantum Gravity, ed. D. Oriti, (Cambridge University Press 2008) contains reviews of different schools of thoughts on quantum gravity.
A nice survey is given by Claus Kiefer, *Quantum Gravity* (Oxford University Press, 2004). A popular account is in *Three Roads to Quantum Gravity* by Lee Smolin (Weidenfeld & Nicholson, London 2000)

Superstring Theory:
M. B. Green, J. H. Schwarz and E. Witten, *Superstring Theory,* (Cambridge University, Cambridge, 1990).
J. Polchinsky, *Superstring Theory* (Cambridge University Press, Cambridge 1998).
For a popular account, read B. Green, *The Elegant Universe*, (Norton & Co. 1999). Or if you prefer a bit more hype:
http://www.pbs.org/wgbh/nova/elegant/
http://superstringtheory.com/people/index.html.
For a critique of this approach, read, e.g., Lee Smolin, *The Trouble With Physics* (Houghton Mifflin, 2006)

Loop Quantum Gravity:
C. Rovelli, *Quantum Gravity* (Cambridge University Press, Cambridge, 2004).
R. Gambini and J. Pullin, *Loops, knots, gauge theories and quantum gravity*, (Cambridge University Press, Cambridge, 1996).
A. Ashtekar and J. Lewandowski, "Background independent quantum gravity: A status report", Class. Quant. Grav. 21 (2004) R53. [gr-qc/0404018]
T. Thiemann, *Introduction to Modern Canonical Quantum General Relativity* (Cambridge University Press, Cambridge, 2008) [gr-qc/0110034]. See also Hermann Nicolai, Kasper Peeters and Marija Zamaklar, "Loop quantum gravity: An outside view" [hep-th/0501114]

Quantum Gravity in 2+1 Dimensions Steven Carlip (Cambridge University Press, 2003)

Simplicial Quantum Gravity,
R. Loll, J. Ambjorn, J. Jurkiewicz, "The Universe from Scratch" Contemp.Phys. **47** (2006) 103-117 [hep-th/0509010]
J. Ambjorn, J. Jurkiewicz, R. Loll, "Quantum Gravity, or The Art of Building Spacetime" [hep-th/0604212] in D. Oriti's book referred to above.

Stochastic Gravity
B. L. Hu, "Stochastic Gravity", Int. J. Theor. Phys. **38**, (1999) 2987 [gr-qc/9902064].
R. Martin and E. Verdaguer, Phys. Rev. D {\bf 60}, 084008 (1999).
B. L. Hu and E. Verdaguer, Class. Quant. Grav. **20** (2003) R1-R42. [gr-qc/0211090]; Living Reviews in Relativity **7** (2004) 3 [gr-qc/0307032].

A Condensed Matter Viewpoint towards the Microstructures of Spacetime:
G. E. Volovik: *The Universe in a Helium Droplet* (Clarendon Press 2003) http://boojum.hut.fi/personnel/THEORY/volovik.html
"Fermi-point scenario for emergent gravity", in Trieste workshop "From Quantum to Emergent Gravity: Theory and Phenomenology" [arXiv:0709.1258] X.- G. Wen, *Quantum Field Theory of Many-Body Systems,* (Oxford University Press 2004) .
C. Barcelo, S. Liberati and M. Visser, *Liv. Rev. Rel.* **8** (2005) 12.
S. Liberati *et al*, Phys. Rev. Lett. 96, 151301 (2006).

Causal Sets and Discrete Structures
R. Sorkin, in *Lectures on Quantum Gravity*, ed. A. Gomberoff and D. Marolf (Springer, N. Y. 2005). Int. J. Theor. Phys. 36, 2759 (1997).
J. Samuel and S. Sinha, Phys. Rev. Lett. **97,** 161302 (2006).

COLLIDING BRANES AND ITS APPLICATION TO STRING COSMOLOGY

YU-ICHI TAKAMIZU[1]

Collaborators: KEI-ICHI MAEDA[1], HIDEAKI KUDOH[3] and GARY GIBBONS[2]

[1] *Department of Physics, Waseda University, Okubo 3-4-1, Shinjuku, Tokyo 169-8555, Japan*

[2] *DAMTP, Centre for Mathematical Sciences, University of Cambridge, Wilberforce Road, Cambridge CB3 OWA, UK*

[3] *Department of Physics, UCSB, Santa Barbara, CA 93106, US*

E-mail: takamizu@gravity.phys.waseda.ac.jp

It is known from the modern accurate observations, surprising us greatly, that the current universe is mainly about 90 percents fulfilled with the unknown dark components: dark energy and dark matter. The most recent WMAP observations are consistent with the universe made up of 74% dark energy, 22% dark matter, and 4% ordinary matter such as baryon. The modern universe not only needs the dark components but also inflaton leading to early accelerating expansion. These unknown components of the universe may be not independent of the unified theory of all forces existing in nature. For the recent studies of the unified theory, one of most promising approaches is a superstring theory, or M-theory, considered as a quantization of fields including gravitational interaction. The cosmological implications of string theory are currently receiving considerable attention, the so-called *string cosmology*. This interest has been inspired in part by the recent advances that have been made towards a non–perturbative formulation of the theory. The goal of superstring cosmology is to examine the dynamical evolution in these theories and re-examine cosmological questions in the light of our new understanding of string theory such as dark energy and dark matter. String theory has a much richer set of fundamental degree of freedom, consisting of D-branes. This fundamental objects, D-branes denote non–perturbative effects of string theory as "soliton" of strings, while string theory has been only described in perturbative form. Inspired by such speculation, recently a new paradigm on the early universe has been proposed, the so-called brane-world. The existence of models with more than one brane suggests that branes may collide. Colliding branes would be a fundamental phenomena in the string cosmology. We have studied several applications of colliding branes to string cosmology.

Keywords: Cosmology; gravity; brane; string theory; colliding branes.

1. Basic Idea of Colliding Branes

It may be difficult to deal properly with the collision of two branes in basic string theory. Hence, in this work, we adopt a domain wall constructed by some scalar field as a brane, and analyze the collision of two domain walls in a 5-dimensional (5D) bulk spacetime. Some other studies have also adopted such a picture. It is worth noting that there is a thick domain wall model for a brane world.[3] In order to analyze particle creation at the brane collision, we consider the simplest situation. We discuss the collision of two domain walls collide in 5D Minkowski spacetime. To construct a domain wall structure, we adopt a real scalar field Φ with a double-well potential, $V(\Phi) = \frac{\lambda}{4}(\Phi^2 - \eta^2)^2$, where the potential minima are located at $\Phi = \pm\eta$. Since we discuss the collision of two parallel domain walls, the scalar field is assumed to depend only on a time coordinate t and one spatial coordinate z. The remaining three spatial coordinates are denoted by $\boldsymbol{x}$. For numerical analysis, we use dimensionless parameters and variables, which are rescaled by η (or its mass scale $m_\eta = \eta^{2/3}$). The equation of motion for Φ in 5D is given by $\ddot{\Phi} - \Phi'' + \lambda\,\Phi\,(\Phi^2 - 1) = 0$, where $\dot{}$ and $'$ denote $\partial/\partial t$ and $\partial/\partial z$, respectively. The basic equation has a static kink solution (K), which is topologically stable. It is called a domain wall, which is described by $\Phi_K(z) = \tanh\left[\frac{z}{D}\right]$ where $D = \sqrt{2/\lambda}$ is the thickness of the wall. We also find another stable solution, that is, the antikink solution ($\bar{K}$) by reflecting the spatial coordinate z as $\Phi_{\bar{K}}(z) = \Phi_K(-z) = -\Phi_K(z)$. When a domain wall moves with constant speed v in the z direction, we obtain corresponding solution as $\Phi_v(z,t) = \tanh\left[\frac{\gamma}{D}(z - vt)\right]$, where we assume that the domain wall is initially located at $z = 0$, and $\gamma = 1/\sqrt{1 - v^2}$ is the Lorentz factor. In order to discuss the collision of two domain walls, we first have to set up the initial data. Provide a kink solution at $z = -z_0$ and an antikink solution at $z = z_0$, which are separated by a large distance and approaching each other with the same speed v. We then obtain the explicit profile; $\Phi(z,0) = \Phi_v(z + z_0, 0) - \Phi_{-v}(z - z_0, 0) - 1$. The initial value of $\dot{\Phi}$ is also given by its derivative. The spatial separation between two walls is given by $2z_0$, and as long as the separation distance is much larger than the thickness of the wall ($z_0 \gg D$), the above initial conditions and give a good approximation for two moving domain walls. Using these initial values, we solve the dynamical equation numerically.

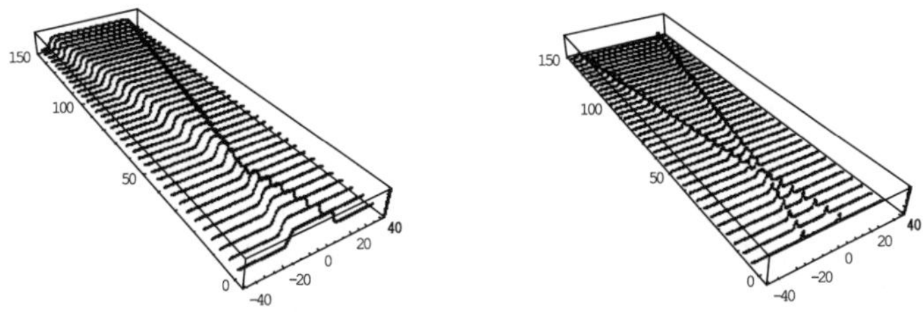

(a) Scalar field (b) Energy density

Fig. 1. Collision of two domain walls where the initial velocity $v = 0.4$. The time evolutions of the scalar field Φ (a) and energy density ρ_Φ (b) are shown from $t = 0$ to 150. The collision occurs once around $t = 31$. We set $\lambda = 1.0$. (b): The maximum point of ρ_Φ defines the position of a wall ($z = z_W(t)$).

2. Reheating Mechanism in Colliding Two Branes Universe

Even though there are some works, the reheating mechanism itself in this scenario has so far not been investigated in detail. Hence, in this section, we study how we can recover the hot big bang universe after the collision of the branes, following our work.[9] This may provide the reheating mechanism of an ekpyrotic brane universe.[6,7] Here we investigate quantum creation of particles, which are confined to the brane, at the collision of two branes. We discuss the collision of two domain walls collide in 5D Minkowski spacetime. We use the unit of $c = \hbar = 1$. We use a numerical approach to solve the equations for the colliding domain walls by using two free parameters, i.e. a wall thickness $D = \sqrt{2/\lambda}$ and an initial wall velocity v. The collision of two walls has been discussed[1] and we find the same results as there. The results are very sensitive to the initial velocity v. First let us show the numerical results for two typical initial velocities, i.e. $v = 0.2$ and 0.4. For the case where $v = 0.4$, the evolution of Φ is depicted in Fig. 1(a), while that of the energy density is shown in Fig. 1(b). From Fig. 1(b), we find some peaks in the energy density, by which we define the positions of moving walls ($z = \pm z_W(t)$). We find the behavior of the collision as follows. Where the initial velocity $v = 0.4$, the collision occurs once, while it does twice where $v = 0.2$. To be precise, in the latter case, after two walls collide, they bounce, recede to a finite distance, and then return to collide again. As shown by the previous work,[1] however, the result highly depends on the incident velocity v. If we change the incident velocity slightly, the number of bounces changes drastically.

Once we find the solution of colliding domain walls, we can evaluate the time evolution of a scalar field on the domain wall. Since we assume that we are living on one domain wall, we are interested in production of a particle confined to the domain wall as a 4D scalar field ψ. The basic equation has a potential term as $\Phi_W(\tau) = \Phi(z_W(\tau))$. It is easy to quantize the scalar field ψ because our background spacetime is 4D Minkowski space by a canonical quantization scheme. After the collision of domain walls, the annihilation and creation operators are given as $\bar{a}_k = \alpha_k a_k + \beta_k^* a_k^\dagger$ where α_k and β_k are the Bogolubov coefficients. As a result, the number density and energy density of produced particles are given by $n = \int |\beta_k|^2 d^3\boldsymbol{k}$ and $\rho = \int |\beta_k|^2 \omega_k d^3\boldsymbol{k}$. Now we estimate the particle production by the domain wall collision. Using the evolution of the scalar field Φ_W, we calculate the Bogolubov coefficients α_k and β_k. We can summarize our results by the following empirical formula $n \approx 25D\bar{g}^4 N_b$ and $\rho \approx 20\bar{g}^4 N_b$, where N_b denotes a number of bounces. If the energy of the particles is thermalized by interaction and a thermal equilibrium state is realized, we can estimate the reheating temperature by $\rho = \frac{\pi^2}{30} g_{\text{eff}} T_{\text{R}}^4$ where g_{eff} is the effective number of degrees of freedom of particles. Hence we find the reheating temperature by the domain wall collision as

$$T_{\text{R}} = \left(\frac{\pi^2}{30}\right)^{-1/4} g_{\text{eff}}^{-1/4} \rho^{1/4} \approx 0.88 \times \left(\frac{g_{\text{eff}}}{100}\right)^{-1/4} \bar{g}\, N_b^{1/4} \,. \tag{1}$$

In order to produce a successful reheating, a reheating temperature must be higher than 10^2 GeV (electro-weak). We find a constraint on the fundamental energy scale m_η as sightly larger than TeV scale.

3. Fermions on Colliding Brane

In the brane-world,[5,8] the idea is that the fermionic chiral matter making up the standard model is composed of such trapped zero modes. Our world is localized on one brane and a shadow world is localized on the other brane. The existence of models with more than one brane suggests that branes may collide, and it is natural to suppose that the Big Bang is associated with the collision.[6,7] This raises the fascinating questions of what happens to the localized fermions during such collisions? In this section, we shall embark on what we believe is the first study of this question by solving numerically the Dirac equation for a fermion coupled via Yukawa interaction to a system of two colliding domain walls in 5D Minkowski spacetime. The back reaction of the fermions on the domain wall is here neglected.

The 5D Dirac equation with a Yukawa coupling term $g\Phi\bar{\Psi}\Psi$ is given by

$$(\Gamma^{\hat{A}}\partial_{\hat{A}} + g\Phi)\Psi = 0, \quad (\hat{A} = 0, 1, 2, 3, 5)\,, \tag{2}$$

where Ψ is a 5D four-component fermion. $\Gamma^{\hat{A}}$ are the Dirac matrices in 5D Minkowski spacetime satisfying the anticommutation relations. Note that

Eq. (2) implies current conservation law as $\partial_A n^A = 0$, where $n^A \equiv \bar{\Psi}\Gamma^{\hat{A}}\Psi$ is conserved number current, which gives conserved number density. Before going to analyze concrete examples, we introduce two chiral fermion states $\Psi_\pm = \frac{1}{2}\left(1 \pm \Gamma^{\hat{5}}\right)\Psi$. As for a fermion, in the case of a static domain wall and assuming massless chiral fermions on a brane, we find the 5D profile of wave function as $f_\pm \propto \left[\cosh\left(\frac{z}{D}\right)\right]^{\mp\epsilon g D}$, where $\epsilon = \pm$ correspond to a kink (anti-kink) solution. Hence the positive-chiral (the negative-chiral) fermion is localized for a kink (an anti-kink) but is not localized for an anti-kink (a kink). To quantize the fermion fields, we define annihilation operators of localized fermions on a kink and on an anti-kink by $a_{\rm K} = \langle \Psi^{(\rm K)}, \Psi\rangle$ and $a_{\rm A} = \langle \Psi^{(\rm A)}, \Psi\rangle$. In what follows, we shall consider only low energy fermions, that is, we assume that $\vec{k} \approx 0$, that is $|\vec{k}|$ is enough small compared with the mass scale of 5D fermion ($g\Phi$). In this case, since up- and down-components are decoupled, we discuss only up-components here. Note that taking into account $\vec{k}$ mixes the up- and down-components. We shall discuss two cases: case (a) collision of two fermion walls, $|{\rm KA}\rangle \equiv a^\dagger_{\rm A} a^\dagger_{\rm K}|0\rangle$ and case (b) collision of fermion and vacuum walls, $|{\rm K0}\rangle \equiv a^\dagger_{\rm K}|0\rangle$. After collision, we find the relations between initial and final states by solving the Dirac equation. We find annihilation operators of final fermion states as

$$b_{\rm K} = \alpha_{\rm K} a_{\rm K} + \beta_{\rm A} a_{\rm A}\,, \quad b_{\rm A} = \alpha_{\rm A} a_{\rm A} + \beta_{\rm K} a_{\rm K}\,. \tag{3}$$

Using the Bogoliubov coefficients $\alpha_{\rm K}, \beta_{\rm K}$ and $\alpha_{\rm A}, \beta_{\rm A}$, we obtain the expectation values of fermion number on a kink and an anti-kink after collision as $\langle N_{\rm K}\rangle \equiv \langle {\rm KA}|b^\dagger_{\rm K} b_{\rm K}|{\rm KA}\rangle = |\alpha_{\rm K}|^2 + |\beta_{\rm A}|^2$ and $\langle N_{\rm A}\rangle \equiv \langle {\rm KA}|b^\dagger_{\rm A} b_{\rm A}|{\rm KA}\rangle = |\alpha_{\rm A}|^2 + |\beta_{\rm K}|^2$ for the case of $|{\rm KA}\rangle$. If the initial state is $|{\rm K0}\rangle$, we find $\langle N_{\rm K}\rangle \equiv \langle {\rm K0}|b^\dagger_{\rm K} b_{\rm K}|{\rm K0}\rangle = |\alpha_{\rm K}|^2$ and $\langle N_{\rm A}\rangle \equiv \langle {\rm K0}|b^\dagger_{\rm A} b_{\rm A}|{\rm K0}\rangle = |\beta_{\rm K}|^2$.

In order to obtain the Bogoliubov coefficients, we have to solve the equations for domain wall Φ^9 and fermion Ψ numerically. Hence, in this paper, we analyze for the case of $g \geq 2$. We set $D = 1$, but leave v free. To obtain the Bogoliubov coefficients, we solve the Dirac equation for the case (b) collision of fermion-vacuum walls. Because of z-reflection symmetry discussed, we find the same Bogoliubov coefficients as $|\alpha_{\rm K}|^2 = |\alpha_{\rm A}|^2$ and $|\beta_{\rm K}|^2 = |\beta_{\rm A}|^2$. The Bogoliubov coefficients depend on the initial wall velocity. We also find that the Bogoliubov coefficients depend sensitively on the coupling constant g as well as the velocity v. We can evaluate the expectation values of fermion numbers after collision as follows. In the case (a), we find that most fermions on domain walls remain on both walls even after the collision. A small amount of fermions escapes into the bulk spacetime at collision. In the case (b), however, we obtain $\langle N_{\rm K}\rangle = |\alpha_{\rm K}|^2$ and $\langle N_{\rm A}\rangle = |\beta_{\rm K}|^2$. Some fermions jump up to the vacuum brane at collision. The amount of fermions localized on which brane depends sensitively on the incident velocity and the coupling constants g/λ where g and λ are the Yukawa coupling constant and that of the double-well potential, respectively. It can be intuitively understood that the amount of localized fermions is roughly determined by the duration of collision for which they transfer to another wall or stay on the initial wall, and the localization condition depending on the Yukawa coupling constant. On the other hands, in the case (a), we find that most fermions seem to stay on both branes even after collision. This is because of the relationship $|\beta_{\rm K}| = |\beta_{\rm A}|$, which is guaranteed by a left-right symmetry in the present system. This result means physically that the same state ($k \sim 0$) of fermions are exchanged for each other by the same amount. Therefore, the final amounts of fermions does not depend on parameters.

4. Collision of Two Domain Walls in Asymptotic Anti de Sitter Spacetime

In order to study whether the reheating process by colliding branes is still efficient in more reliable cosmological models, we have to include the curvature (self-gravity) effect. In particular, some brane universe are discussed with a negative cosmological constant[8] Hence, we study here how gravitational effects change our previous results. Inspired by the Randall Sundrum brane model,[8] we include a potential of the scalar field which provides an effective negative cosmological constant in a bulk spacetime. We solve the 5D Einstein equations and the dynamical equation for a scalar field to analyze collision of thick walls in asymptotically anti-de Sitter (AdS) spacetime. It is possible to choose coordinates such that the bulk metric has the "2D conformal gauge" $ds^2 = e^{2A(t,z)}(\,-dt^2 + dz^2\,) + e^{2B(t,z)}d\boldsymbol{x}^2$. This gauge choice also makes the initial setting easy when we construct moving domain walls by use of the Lorentz boost. For an initial configuration of a domain wall, we use an exact static solution.[3] They assume a scalar field Φ with a potential $V(\Phi) = \left(\frac{\partial W}{\partial \Phi}\right)^2 - \frac{8}{3}\kappa_5^2\, W^2$, where $W \equiv \frac{1}{D}\left(\Phi - \frac{1}{3}\Phi^3 - \frac{2}{3}\right)$ is a superpotential, and κ_5^2 and D are the five dimensional gravitational constant and the thickness of a domain wall, respectively. With this potential, we can obtain the metric of 5D spacetime given as $ds^2 = e^{2A_K(y)}(-dt^2 + d\boldsymbol{x}^2) + dy^2$.[3] This metric approaches that of the AdS spacetime in one asymptotic region ($y \ll -1$), i.e., $e^{2A_K} \to e^{-2k|y|}$ as $y \to -\infty$, with $k = \frac{8\kappa_5^2}{9D}$. While it becomes

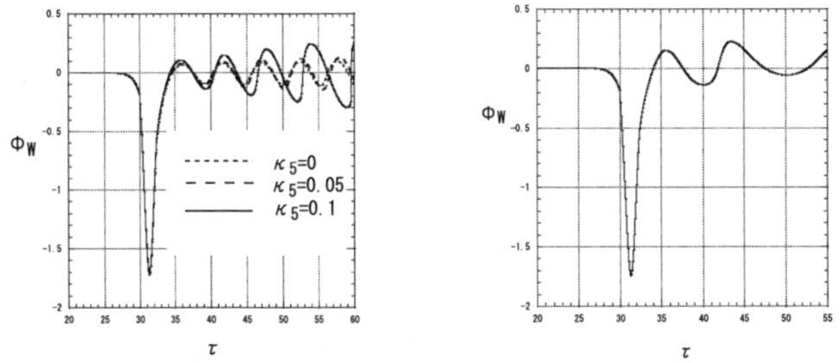

(a) $\kappa_5 \leq 0.1$ (b) $\kappa_5 = 0.15$

Fig. 2. Time evolution of a scalar field on the moving wall with $v = 0.4$, $D = \sqrt{2}$. We set $\kappa_5 = 0, 0.05, 0.1$ for (a) and $\kappa_5 = 0.15$ for (b). The value of the scalar field on the wall is defined by $\Phi_W = \Phi(t, z_W(t))$, where $z_W(t)$ is the position of the wall. (a): The time of period of oscillation gets slightly longer and the amplitude is a little bit larger as κ_5 increases. (b): Φ_W goes out of oscillation phase 0 at $\tau \sim 55$ and then numerical simulation stops.

a flat Minkowski space in another asymptotic region ($y \gg 1$). By reflecting the spatial coordinate y, we also find an antikink solution ($\bar{K}$). In order to describe this solution under our gauge condition in the (t, z) frame, we should transform the present (t, y) coordinates into the (t, z) ones by defining the coordinate z as $z = \int e^{-A_K(y)} dy$. We adopt a numerical method similar to one used in.[9] In what follows, fixing the values as $D = \sqrt{2}$ and $v = 0.4$, we show our results for changing a warp factor k (or gravitational constant κ_5). Once we find the solution of colliding domain walls, we know the time evolution of a scalar field on the domain wall, and we can evaluate production rate. At the position of a domain wall, we can obtain the proper time and the scale factor determined as $\tau = \int e^{A_W} dt$ and $a(\tau) = e^{B_W}(\tau)$, where A_W and B_W are evaluated on the brane. We depict the time evolution of Φ_W on one moving wall for different values of κ_5 in Figs. 2(a) and 2(b). The feature of collision is similar, but the behaviour of a scalar field on the moving wall after collision is different for each κ_5. For a small value of κ_5, e.g., $\kappa_5 = 0.01$, the result is almost the same as the case of the Minkowski background, in which case we find one bounce point, and then the oscillations around $\Phi_W = 0$ follow (see the dotted line in Fig. 2(a)). This oscillation is explained by using a perturbation analysis in Minkowski spacetime.[9] We have found one stable oscillation mode around the kink solution. This oscillation appears by excitation of the system at collision. For $\kappa_5 \geq 0.15$, the behaviour of this oscillation changes drastically. After several oscillations, the scalar field leaves $\Phi_W = 0$ as shown in Fig. 2(b). The numerical simulation eventually breaks down because all variables diverge. The metric component A at $z = 0$ also diverges. It is not a coordinate singularity, but a curvature singularity since we can find the Kretschmann invariant scalar of $R^{abcd}R_{abcd}$ also diverges and the constraint equations are always satisfied. From our analysis, we conclude that gravitational back reaction makes a kink solution unstable contrary to the Minkowski case. We evaluate the time evolutions of the metric $A_{\rm W}, B_{\rm W}$ on the brane. From those two quantities, we evaluate a scale factor of our universe and the Hubble expansion parameter. We find that our universe expands slightly before bounce then eventually contracts, which is inconsistent with the hot big bang universe.

5. Colliding Branes and Black Brane Production

It is known that primordial black holes and domain walls may have been produced in the early universe through the physical process of the collapse of cosmological density perturbations and the series of phase transitions during the cooling phase of universe. On the other hand, black holes and domain walls (also known as branes) also play an important role in string theory as fundamental constituents. In view of the phenomenological relevance, understanding how the domain walls/branes interact dynamically is an important problem, and more knowledge in this area could help in clarifying many issues regarding the early universe. In the past few years much attention have been paid to understanding the dynamics of domain walls. In particular, the interaction between black holes and domain walls has been the subject of study. Nevertheless, even more fundamental processes like collision, recoil, and reconnection of branes are less understood. In this section, we consider the problem from a different perspective. We investigate the process of collision using a BPS domain wall in 5D supergravity, and our main goal is to determine the final outcome of the kink-anti-kink collisions including self-gravity. The system we intend to study consists of two domain walls that are initially located far away from each other. The solution of 5D Einstein equations represent a single domain wall, which is identical to the exact BPS solution in 5D supergravity.[2] There are three unfixed parameters, i.e., wall thickness D, amplitude L of scalar field. The scalar field asymptotes constants, and the scalar potential plays the role of the cosmological constant $\Lambda = \kappa^2 V$ in the limit $\Lambda = -\frac{16L^4}{3D^2}$. It becomes the wall solutions interpolating between AdS and flat Minkowski vacua. We shall restrict our analysis to collisions along a y-direction, preserving the symmetry along the homogeneous $\vec{x}_3$-directions. The initial data for such a collision can be obtained as follows. First of all, we introduce a new coordinate z by $z = \int dye^{-U}$ and work on the conformal gauge. Then the above single static wall is boosted along the fifth direction z, and we obtain a wall moving with constant velocity v.[10] The kink and anti-kink solutions are characterized by their own width and amplitude, (D_K, L_K) and (D_A, L_A), respectively. We have solved the system numerically and evaluated the constraints at each time step for various families of initial data. After collision, the curvature diverges rapidly at the point, whereas the curvature on the wall remains finite and small at the moment. For the wide range of initial parameters, the emergence of singularity is the generic consequence. However, as expected and discussed below, the singularity does not appear for $L \ll 1$ and/or $v_0 \ll 1$ for fixed D. In order to confirm the nature of singularity, we evolve the colliding walls in the double-null coordinates. The numerical result shows that the curvature singularity is spacelike, approaching null lines at late times, which corresponds to the event horizon. This result is very generic, and we have observed similar results for the wide range of initial data (velocity, etc.). This system has homogeneous 3-spatial directions, and so the horizon also extends in these directions. This means that a black brane is produced by the collision of walls, so that this collision provides the dynamical mechanism of generating black branes in higher dimensions.

6. Conclusion

First, we have estimated a reheating temperature by collision, which is related to the ekpyrotic universe scenario (Sec. 2). For simplify, in the case of Minkowski spacetime, we have calculated a quantum particle creation and estimated the reheating temperature as $T_{\rm R} \approx 0.88\ \bar{g}\ N_b^{1/4}$, where $\bar{g}$ and N_b denote the coupling constant and the number of bounces. It can provide an efficient value of reheating temperature to have the baryogenesis at the electro-weak energy scale. Moreover we have considered a standard particles (fermions) which is confined on such domain walls (Sec. 3). We found that most fermions are localized on both branes as a whole even after collision. However, how much fermions are localized on which brane depends sensitively on the incident velocity and the coupling constants unless the fermions exist on both branes. Since we have discussed only the case of zero-momentum fermion on branes ($\vec{k} = 0$), we have only a single state on each brane. If we take into account degree of freedom of low energy fermions, we can put different states of fermions on each brane and can discuss a pair production of fermion and antifermion.

Based on Sec. 4, including self-gravity, we studied collision of two domain walls in 5D asymptotically Anti de Sitter spacetime. We have evaluated a change of scalar field making the domain wall and investigated the effect of a negative cosmological term in the bulk to both the collision process and the evolution of our universe. For large case, it becomes an unstable oscillation and then the singularity appears after collision. In the time evolution of our universe, we find that the universe first expands a little just before collision and then contracts just after collision. We cannot explain our hot big bang universe as it is. However, we have found a possibility to formation of higher dimensional black holes by colliding branes (Sec. 5). After a collision, a spacelike curvature singularity covered by a horizon is formed in the bulk, resulting in a black brane with trapped domain walls. This is a generic consequence of collisions, except for non-relativistic weak field cases. These results show that incorporating the self-gravity drastically changes a naive picture of colliding branes. Finally, for the topic of dimensionality problem, we have studied a thermal equilibrium of string gas at Hagedorn temperature, however, in the paper, we skip it and see the paper[11] for the details. In this work, we found a solution, which implied an interesting possibility for constructing a model that resolves the stabilization and dimensionality problem at the same time.

References

1. P. Anninos, S. Oliveira and R. A. Matzner, Phys. Rev. D **44**, 1147 (1991).
2. M. Arai, S. Fujita, M. Naganuma, and N. Sakai, Phys. Lett. B **556**, 192 (2003).
3. M. Eto and N. Sakai, Phys. Rev. D **68**, 125001 (2003).
4. G.W. Gibbons, K. Maeda and Y. Takamizu, Phys. Lett. B **647**, 1 (2007).
5. P. Horava and E. Witten, Nucl. Phys. B **460**, 506 (1996); Nucl. Phys. B **475**, 94 (1996).
6. J. Khoury, B. A. Ovrut, P. J. Steinhardt and N. Turok, Phys. Rev. D **64**, 123522 (2001).
7. J. Khoury, B. A. Ovrut, N. Seiberg, P. J. Steinhardt and N. Turok, Phys. Rev. D **65**, 086007 (2002).
8. L. Randall and R. Sundrum, Phys. Rev. Lett. **83**, 3370 (1999); Phys. Rev. Lett. **83**, 4690 (1999).
9. Y. Takamizu and K. Maeda, Phys. Rev. D **70**, 123514 (2004).
10. Y. Takamizu and K. Maeda, Phys. Rev. D **73** 103508 (2006).
11. Y. Takamizu and H. Kudoh, Phys. Rev. D **74**, 103511 (2006).
12. Y. Takamizu, H. Kudoh and K. Maeda, Phys. Rev. D **75**, 061304 (2007).

ONE-LOOP CORRECTIONS TO SCALAR AND TENSOR PERTURBATIONS DURING INFLATION

Y. URAKAWA* and K. MAEDA

Science and Engineering, Waseda University,
Tokyo 169-8555, Japan
**E-mail: yuko at gravity.phys.waseda.ac.jp*

Based on the stochastic gravity, we study the loop corrections to the scalar and tensor perturbations during inflation. Since the loop corrections to scalar perturbations suffer the infrared(IR) divergence, we consider the IR regularization to obtain the finite value. We find that the loop correction is amplified by the e-folding, in other words there appears the logarithmic corrections, just as discussed by M.Sloth et.al. On the other hand, we find that the tensor perturbations do not suffer from the infrared divergence.

Keywords: Inflation; stochastic gravity.

1. Introduction

Inflation provides a natural framework which explains both the large-scale homogeneity of the universe and its small-scale irregularity. Despite of its attractive natures, we still have many unknowns about the theory of inflation, since in most models, inflation takes place at energy scale by many orders of magnitude higher than that accessible by accelerators. That is why it is necessary to explore all that we can learn about this high energy regime from the signatures left by inflation in the present universe.[1–4]

However, when we consider the power spectrum of the curvature perturbation ζ only by the linear analysis, many inflation models may predict the same results, which are compatible with the observational data, although the fundamental theories are quiet different. In order to discriminate between different inflationary models, it is important to take into account non-linear effects.[5–16] In particular, the classical perturbation theory predicts that when we consider most of inflation models, the curvature perturbation ζ, which is directly related to the fluctuation of the temperature of CMB, is conserved in superhorizon region.[17–19] In that case, the primordial perturbation is essentially characterized by the behaviour of the background inflaton field near the time of horizon exit. Although this fact makes the computation of the generated primordial perturbations simple, it makes difficult to discriminate different inflation models. That is why the non-local dependence on the evolution of the background scalar field has been studied among the non-linear quantum effects such as the loop corrections.[9,10] However, it is difficult to compute these non-linear quantum effects despite of those importance. This is because they contain integrations about the internal momenta.[11–14] Furthermore, there are several types of non-linear interactions which induce loop corrections, such as self interaction of a scalar field and interaction between matter field and gravitational field. Depending on the interaction term, we find different behaviour of loop corrections.

Stochastic gravity may be well-suited to compute loop corrections which are induced from interactions between a scalar field and gravitational field. Stochastic gravity was proposed in order to discuss the behaviour of the gravitational field in sub-Planck scale, which is affected by quantum matter fields.[20–28] From our naive expectation, on this energy scale, the quantum fluctuation of matter field may dominate that of the gravitational field. Based on this insight, Martin and Verdaguer have presented the evolution equation of the gravitational field, which is affected by a quantum scalar field.[23] The induced effect by the quantum matter field is evaluated by the so-called closed time path (CPT) formalism.[29–33] We integrate the action over quantum scalar fields. As a result the evolution equation of the gravitational field is described by the Langevin type equation, which is called the Einstein-Langevin equation. In our previous work,[39] we have applied their formalism to discuss the evolution of the primordial perturbations, especially, the curvature perturbation ζ, which is important to consider the imprint on observational data. We find that it reproduces the same results as the prediction obtained by the quantization of the gauge invariant variables,[35,36] except for the limited case. Only when the e-folding from the horizon crossing time to the end of inflation exceed some critical value, the Einstein-Langevin equation in[23] does not give the same result as that of the gauge invariant variables. It is because we do not quantized the longitudinal part of the gravitational field which is induced by a quantum scalar field.[23] However, we should stress that stochastic gravity will still provide the correct calculation when the e-folding is smaller than the critical value. In that case, stochastic gravity is well-suited for loop calculations. Hence, in this paper, using stochastic gravity, we evaluate loop corrections induced by quantum scalar fields.

In general, loop corrections contain a divergent part. In a quantum field theory in Minkowski spacetime, the divergence usually appears at the high energy scale. In order to discuss finite and physical quantities, appropriate regularization and renormalization are required. Apart from such an ultraviolet divergence, there may appear another divergence in the de Sitter (or quasi de Sitter) spacetime. It is the infrared divergence. When we cannot neglect the effects of the gravitational field, the definition of the positive frequency mode function becomes ambiguous. The infrared divergence is related to the choice of initial condition for the mode function.

In this paper, we discuss this infrared problem, taking into account the behaviour of one-loop corrections to the scalar and the tensor perturbations. The infrared problem is important because if we introduce an infrared cutoff to obtain a finite value, the final amplitude has a logarithmic correction. Such a logarithmic correction amplifies the loop corrections. We also find that there is a crucial difference between the logarithmic corrections in scalar and tensor perturbations, which is related to the infrared divergence.

In this paper, we consider a minimally coupled single-field inflation as a simple slow-roll inflation model, which action is given by

$$S[g,\,\phi] = \int d^4x\sqrt{-g}\Big[\frac{1}{2\kappa_B^2}(R-2\Lambda_B) + \alpha_B C_{abcd}C^{abcd} + \beta_B R^2 \\ -\frac{1}{2}\{g^{ab}\partial_a\phi\partial_b\phi + 2V(\phi)\}\Big] \quad (1)$$

where $\kappa_B^2 \equiv 8\pi G_B$ is the bare gravitational constant. The subscript "B" represents the values of bare coupling constants. After we regularize divergent parts and renormalize them, we set the renormalized constants as $\alpha = \beta = \Lambda = 0$ for simplicity. We also represent the renormalized gravitational constant by $\kappa^2 \equiv 8\pi G$. In order to characterize the slow-roll inflation, we adopt two slow-roll parameters; $\varepsilon \equiv -\dot{H}/H^2$ and $\eta_V \equiv V_{,\phi\phi}/\kappa^2 V$. As for time variable, we use the conformal time, τ, and represent the time derivative by a prime.

2. Stochastic Gravity

First we shortly summarize the basic points on stochastic gravity. The basic equation in stochastic gravity describes the evolution of the gravitational field, which source term is given by quantum matter fields. The effective action obtained by integrating matter fields with the CTP formalism. We find the Langevin type equation, i.e. the so-called Einstein-Langevin equation which is analogous to the equation of motion for the Brownian particle. In the Brownian motion case, the deterministic motion is influenced and modified by stochastic behaviour of the environmental source. It is worth while noting that this Langevin type equation is well-suited not only to understand the properties of inflation and the origin of large-scale structures in the Universe, but also to explain transition from quantum fluctuations to classical seeds.

When we consider an interacting quantum system, which includes the gravitational field on the sub-Planck scale, we expect that quantum fluctuations of matter fields dominate that of the gravitational field. In stochastic gravity, we assume that the gravitational field is not quantized but can be treated classically because we are interested in a sub-Planckian scale. However, it is important to take into account fluctuation of gravitational field which is induced through interaction with quantum matter fields. In order to discuss such dynamics of the gravitational field, the CTP formalism is useful.

The effective action for in-in expectation value of the gravitational field is derived by integrating matter fields. In Ref.,[23] Martin and Verdaguer derived the effective equation of motion based on the CTP functional technique applied to a system-environment interaction, more specifically, based on the influence functional formalism by Feynman and Vernon. This CTP effective action contains two specific terms, in addition to the ordinary Einstein-Hilbert action, which describe the induced effects through interaction with quantum matter fields. One is a memory term, by which the equation of motion depends on the history of the gravitational field itself. The other is a stochastic source ξ_{ab}, which describes quantum fluctuation of a scalar field. The latter one is obtained from the imaginary part of the effective action, and as such it cannot be interpreted as a conventional action. Indeed, there appear statistically weighted stochastic noises as a source for the gravitational field. Under the Gaussian approximation, this stochastic variable is characterized by the average value and the two-point correlation function;

$$\begin{aligned} &\langle \xi_{ab}(x) \rangle = 0 \,, \\ &\langle \xi_{ab}(x_1) \xi_{c'd'}(x_2) \rangle = N_{abc'd'}(x_1,\ x_2) \,, \end{aligned} \tag{2}$$

where the bi-tensor $N_{abc'd'}(x_1,\ x_2)$ is called a noise kernel, which represents quantum fluctuation of the energy-momentum tensor in a background spacetime, i.e.

$$N_{abc'd'}(x_1,x_2) \equiv \frac{1}{4}\mathrm{Re}[F_{abc'd'}(x_1,x_2)]$$

$$= \frac{1}{8}\langle\{\hat{T}_{ab}(x_1) - \langle\hat{T}_{ab}(x_1)\rangle, \hat{T}_{ab}(x_2) - \langle\hat{T}_{ab}(x_2)\rangle\}\rangle[g]\,, \qquad (3)$$

where $\{\hat{X},\hat{Y}\} = \hat{X}\hat{Y} + \hat{Y}\hat{X}$, g is the metric of a background spacetime, and the bi-tensor $F_{abc'd'}(x,y)$ is defined by

$$F_{abc'd'}(x_1,x_2) \equiv \langle\hat{T}_{ab}(x_1)\hat{T}_{c'd'}(x_2)\rangle[g]$$

$$-\langle\hat{T}_{ab}(x_1)\rangle[g]\ \langle\hat{T}_{c'd'}(x_2)\rangle[g]\,. \qquad (4)$$

The expectation value for the quantum scalar field is evaluated in the background spacetime g.

Including the above-mentioned stochastic source of ξ_{ab}, the effective equation of motion for the gravitational field is written as

$$G^{ab}[g+\delta g] = \kappa^2\left[\langle\hat{T}^{ab}\rangle_R[g+\delta g] + 2\xi^{ab}\right]\,, \qquad (5)$$

where δg is the metric perturbation induced by quantum fluctuation of matter fields and stochastic source ξ_{ab} is characterized by the average value and the two-point correlation function Eq.(2).

Note that this equation is the same as the semiclassical Einstein equation expect for a source term of stochastic variables ξ_{ab}. Furthermore, the expectation value of energy-momentum tensor includes a non-local effect as follows. It consists of three terms as

$$\langle\hat{T}^{ab}\rangle_R[g+\delta g] = \langle\hat{T}^{ab}(x)\rangle[g] + \langle\hat{T}^{(1)ab}[\phi[g],\delta g](x)\rangle[g]$$

$$-2\int d^4y\sqrt{-g(y)}H^{abcd}[g](x,y)\delta g_{cd}(y) + O(\delta g^2)\,, \qquad (6)$$

where the expectation value of $\hat{T}^{(1)ab}$ and H^{abcd} are defined in below (Eq.(7) and (9)). The evolution equation for a scalar field depends on the gravitational field. As a result, the expectation value of energy-momentum tensor (Eq.(6)) depends not only directly on the spacetime geometry but also does indirectly through a scalar field. When we perturb a spacetime as $(g+\delta g)$, two different changes appear in the right hand side of Eq.(6). The second term in Eq.(6) represents the direct change, which is described by fluctuation of the gravitational field δg as

$$\langle \hat{T}^{(1)ab}[\phi[g], \delta g](x) \rangle$$

$$= \Big(\frac{1}{2} g^{ab} \delta g_{cd} - \delta^a{}_c g^{be} \delta g_{de} - \delta^b{}_c g^{ae} \delta g_{de} \Big) \langle \hat{T}^{cd} \rangle [g]$$

$$- \Big\{ \Big(1 - \frac{2\,\varepsilon}{3} \Big) \rho + \frac{\delta\psi^2}{2a^2} \Big\} \Big(g^{ac} g^{bd} - \frac{1}{2} g^{ab} g^{cd} \Big) \delta g_{cd} \,. \tag{7}$$

where $\delta\psi^2$ is defined in terms of the quantum fluctuation of the scalar field ψ as follows,

$$\delta\psi^2 \equiv \langle \nabla_0 \psi \nabla_0 \psi + \gamma^{ij} \nabla_i \psi \nabla_j \psi \rangle [g] \tag{8}$$

We can neglect this term safely on the sub-Planck scale, because this term is smaller by order of $(\kappa H)^2$ than the preceding term. To derive this expression, we have used the background evolution equation for a scalar field.

The third integral term in the r.h.s. of Eq.(6) represents the effect from the indirect change and is characterized by the dissipation kernel, which is given by

$$H_{abc'd'}(x_1, x_2) = H^{(\mathrm{S})}_{abc'd'}(x_1, x_2) + H^{(\mathrm{A})}_{abc'd'}(x_1, x_2) \tag{9}$$

$$H^{(\mathrm{S})}_{abc'd'}(x_1, x_2) = \frac{1}{4} \mathrm{Im}[S_{abc'd'}(x_1, x_2)] \tag{10}$$

$$H^{(\mathrm{A})}_{abc'd'}(x_1, x_2) = \frac{1}{4} \mathrm{Im}[F_{abc'd'}(x_1, x_2)] \,, \tag{11}$$

where $S_{abc'd'}(x_1, x_2)$ is defined by

$$S_{abc'd'}(x_1, x_2) \equiv \langle T^* \hat{T}_{ab}(x_1) \hat{T}_{c'd'}(x_2) \rangle [g] \,, \tag{12}$$

T^* denotes that we take time ordering before we apply the derivative operators in the energy momentum tensor. As pointed out in,[23] only if the background spacetime g satisfies the semiclassical Einstein equation, the gauge invariance of the Einstein-Langevin equation is guaranteed. Hence, in this paper, to guarantee the gauge invariance, we assume the background spacetime satisfies the semiclassical Einstein equation.

The Einstein-Langevin equation, (5) contains two different sources. One is a stochastic source ξ_{ab}, whose correlation function is given by the noise kernel. From the explicit form of a noise kernel (3), we find that ξ_{ab} represents the quantum fluctuation of the energy momentum tensor. The other is an expectation value of the energy momentum tensor in the perturbed spacetime $(g + \delta g)$, which includes a memory term. The integrand of a memory term consists of a dissipation kernel and fluctuation of the gravitational field. In order to investigate the evolution for fluctuation of the gravitational field, it is necessary to calculate the quantum correction of a scalar field and evaluate the noise kernel and the dissipation kernel. Note that the noise kernel and the dissipation kernel correspond to the contributions from internal lines or loops of the Feynman diagrams, which consist of propagators of a scalar field and do not include that of the gravitational field in our approach.

3. Loop Corrections to the Correlation Functions

In this section, we consider the loop corrections to the scalar perturbations and the tensor perturbations. Here we adopt the following metric form:

$$\begin{aligned} ds^2 &= -a^2(\tau)(1+2\mathcal{A}_{\boldsymbol{k}}Y_{\boldsymbol{k}})d\tau^2 - 2a^2(\tau)\frac{k}{\mathcal{H}}\Phi_{\boldsymbol{k}}Y_{j\boldsymbol{k}}d\tau dx^j \\ &\quad + a^2(\tau)\left(\gamma_{ij} + 2H^{(t)}_{T\boldsymbol{k}}e_{ij}(\boldsymbol{k})Y_{\boldsymbol{k}}dx^i dx^j\right. \end{aligned} \tag{13}$$

where a and γ_{ij} are the scale factor and the metric of maximally symmetric three space. The scalar perturbations are described by $\mathcal{A}$ and $(k/\mathcal{H})\Phi$, which are the so-called lapse function and shift vector, respectively. $H^{(t)}_T$ is the tensor perturbation. The scalar perturbations and the tensor perturbations are expanded by a complete set of harmonic function. Since this gauge choice fixes both the time slicing and the spatial coordinate completely, all physical variables with this ansatz are gauge invariant. This choice of the time coordinate is called a flat slicing, because the spatial curvature vanishes in this slicing.

3.1. *Scalar Perturbations*

Among the scalar perturbations, we consider the curvature perturbation in uniform density slicing, ζ. The curvature perturbation ζ is proportional to the gravitational potential in the late time of the universe and it is directly related to the fluctuation of the temperature of CMB. That is why it is important for us to consider this gauge-invariant variable among scalar perturbations. This curvature perturbation ζ is related to density perturbation in flat slicing δ_f by

$$\zeta = \frac{1}{2\varepsilon}\delta_f \tag{14}$$

To evaluate the correlation function of the density perturbation in flat-slicing δ_f, focusing on the proper non-linear effects, we neglect the contribution from the product of linear perturbations.

Solving the Einstein Langevin equation, we find the correlation function of the density perturbation δ_f as

$$\langle \delta_{f\boldsymbol{k}}(\tau)\delta_{f\boldsymbol{p}}(\tau)\rangle^{(4)}$$

$$\simeq \frac{\pi}{2}\frac{(\kappa H_k)^4}{k^3}\eta_V^2\delta(\boldsymbol{k}+\boldsymbol{p})\left\{\frac{1}{3}-\frac{1-(k/H_i)^{2(\varepsilon-\eta_V)}}{2(\varepsilon-\eta_V)}\right\} \tag{15}$$

Taking into account Eq.(14), we obtain the loop corrections to the correlation function of the curvature perturbation in uniform density slicing ζ as

$$\langle \zeta_{\boldsymbol{k}}(\tau)\zeta_{\boldsymbol{p}}(\tau)\rangle^{(4)}$$

$$\simeq \frac{\pi}{8}\frac{(\kappa H_k)^4}{k^3}\left(\frac{\eta_V}{\varepsilon}\right)^2\delta(\boldsymbol{k}+\boldsymbol{p})\left\{\frac{1}{3}-\frac{1-(k/H_i)^{2(\varepsilon-\eta_V)}}{2(\varepsilon-\eta_V)}\right\}. \tag{16}$$

The final result depends on the initial Hubble Horizon scale, $\mathcal{H}_i = H_i$, which is introduced to remove the infrared divergence. The case with $2|\varepsilon-\eta_V|\log(k/\mathcal{H}_i) < 1$ is particularly interesting. This, in the other words, corresponds to the case of $N_k < 1/2|\varepsilon-\eta_V|$, where $N_k \simeq \log(k/\mathcal{H}_i)$ is the e-folding from the beginning of inflation to the horizon crossing time. In this case, this correlation function is approximated as

$$\langle \zeta_{\boldsymbol{k}}(\tau)\zeta_{\boldsymbol{p}}(\tau)\rangle^{(4)}$$

$$\simeq \frac{\pi}{8}\frac{(\kappa H_k)^4}{k^3}\left(\frac{\eta_V}{\varepsilon}\right)^2\delta(\boldsymbol{k}+\boldsymbol{p})\left(\frac{1}{3}+N_k\right) \tag{17}$$

Note that there appears the logarithmic corrections. This results say that although the one-loop correction is suppressed by $(\kappa H_k)^4$ and is smaller by order of $(\kappa H_k)^2$ than tree level effects, it is amplified by the e-folding N_k from the initial time to the horizon crossing time, which can become large contrary to the e-folding from the horizon crossing time to the end of the inflation. However note that this amplification is derived from the infrared divergence. In Sec.4, we discuss the origin of the infrared divergence and the possibility of such an amplification.

3.2. *Tensor Perturbations*

Similar to the scalar perturbations, the tensor part of the Einstein-Langevin equation implies that the loop corrections to the correlation function of the tensor perturbations becomes

$$\langle H^{(t)}_{T\boldsymbol{k}}(\tau)e^{i}{}_{j}(\boldsymbol{k})\; H^{(t)}_{T\boldsymbol{p}}(\tau)e^{j}{}_{i}(\boldsymbol{p})\rangle^{(4)}$$

$$\simeq \frac{\pi}{135}\frac{(\kappa H_k)^4}{k^3}\;\delta(\boldsymbol{k}+\boldsymbol{p})\left(\frac{7}{6}-15\,x^{2+2\eta_V}\right) \tag{18}$$

It is interesting to note that there is no dependence on the infrared cut off H_i in the tensor perturbations. Furthermore, the tensor perturbations is divergent free in superhorizon region, as was pointed out in our previous work.[39] We shall discuss the reason in the next section.

4. Discussion

Despite its attractive nature, there would be the large ambiguity about inflation models. In order to clarify its fundamental properties, it is important to analyze not only the linear perturbations, but also the non-linear one. That is why the interests about the non-linear perturbations are gradually increasing in recent days. Among them, we have studied one-loop corrections to the scalar perturbations ζ and the tensor perturbations in the formalism of stochastic gravity. In stochastic gravity, based on the naive expectation that quantum fluctuation of the gravitational field is sufficiently small at the sub-Planck scale, we assume that contribution from the Feynman diagrams where the gravitational field propagates as an internal line is negligibly small. When we compute the CTP effective action, we integrate out only the dynamical degree of freedom of a scalar field. As a result, from this coarse-grained effective action, the Einstein-Langevin equation for the gravitational field is derived. Using this Langevin type equation, we can discuss the loop corrections to the scalar perturbations and the tensor perturbations, which are amplified through the non-linear interaction between the scalar field and the gravitational field.

When we consider the loop corrections in inflationary universe, there are two different divergences. One is the ultraviolet (UV) divergence. Since this divergence is originated by short wave modes, such divergence also appear in the quantum field theory in a Minkowski background. In inflationary spacetime, there exists another divergence which is not found in a Minkowski spacetime. It is the IR divergence. It is necessary to take into account the effects of the curvature for long wave modes beyond the Hubble horizon scale from the beginning of inflation. Once the gravity is included, it becomes difficult to choose the physical positive frequency mode function, because the mode function is modified non-trivially by the coordinate transformations. When we impose the same initial condition on those long wave modes as the short wave modes, the loop integral diverges on the IR side. In order to avoid this IR divergence, we have introduced the cut-off at the initial Hubble horizon size. Then the amplitude of the one-loop corrections to the curvature perturbation ζ is amplified by the e-folding from the initial time of inflation to the horizon crossing time, i.e. the logarithmic correction. If it is true, this amplification may make it possible to detect these loop corrections. Then it will be a great help to clarify the fundamental properties of an inflation model.

The notable fact is that, although the scalar perturbations are amplified by the logarithmic corrections, the tensor perturbations are not. Even if we remove the IR cut-off, the IR divergence does not appear in tensor perturbations. This difference between the scalar perturbations and the tensor perturbations seems to be related to the origin of this logarithmic corrections.

In order to consider the origin of this logarithmic corrections due to the IR cut-off, we first mention about the prediction in stochastic inflation.[34] In stochastic inflation, the long wave modes of the scalar field couple to the short wave modes through the self interaction of the scalar field. Then the long wave modes are affected by the quantum fluctuation of the short wave modes. As a result, the long wave modes come to show the stochastic behaviour. This stochastic behaviour of the long wave modes affect the background quantities.

In our case, due to the non-linear interaction between the gravitational field and the scalar field, the long wave modes and the background quantities come to show the stochastic behaviour. Since the scalar perturbations are defined as the deviation from the background quantities, the stochastic fluctuation of the background quantities affect the behaviour of the perturbed variables. As a result, it induces the logarithmic secular evolution of the perturbed variables. On the other hand, there are no background tensor mode, then the tensor perturbation can avoid to be affected from the background stochastic fluctuations.

Furthermore, although, in this paper, as a simplest step to treat the IR divergence, we have just neglected the long wave modes with $-k\mathcal{H}_i > 1$, the more careful treatment is required about the infrared modes. We will postpone it as the future work, too.

There is another notable difference between the scalar perturbations and the tensor perturbations. As pointed out in our previous work, in the present approach of stochastic gravity the longitudinal part of the quantized gravitational field is not included. It affects the behaviour of perturbations in superhorizon region. In particular, the curvature perturbation deviates from constant when the e-folding from the horizon crossing time exceeds the definite critical value ($|\text{slow-roll parameter}|^{-1}$). Since this is the problem at the level of the propagator, the loop corrections to the scalar perturbations are also influenced by the non-existence of the longitudinal part of the quantized gravitational field. In fact, we can show that the one loop corrections to the curvature perturbation ζ evolves as $x^{4\eta_V}$ in super horizon region. On the other hand, as shown in Eq.(18), the tensor perturbations does not decay. It means that even if we neglect the longitudinal part of the gravitational field, it does not affect to the one loop corrections to the tensor perturbations.

References

1. J. E. Lidsey, A. R. Liddle, E. W. Kolb, E. J. Copeland, T. Barreiro and M. Abney, Rev. Mod. Phys. **69**, 373 (1997) [arXiv:astro-ph/9508078].
2. B. A. Bassett, S. Tsujikawa and D. Wands, Rev. Mod. Phys. **78**, 537 (2006) [arXiv:astro-ph/0507632].
3. D. H. Lyth, arXiv:hep-th/0702128.
4. A. Linde, arXiv:0705.0164 [hep-th].
5. N. Bartolo, E. Komatsu, S. Matarrese and A. Riotto, Phys. Rept. **402**, 103 (2004) [arXiv:astro-ph/0406398].
6. J. M. Maldacena, JHEP **0305**, 013 (2003) [arXiv:astro-ph/0210603].
7. D. Seery and J. E. Lidsey, JCAP **0506**, 003 (2005) [arXiv:astro-ph/0503692].
8. D. Seery and J. E. Lidsey, JCAP **0509**, 011 (2005) [arXiv:astro-ph/0506056].
9. S. Weinberg, Phys. Rev. D **72**, 043514 (2005) [arXiv:hep-th/0506236].
10. S. Weinberg, Phys. Rev. D **74**, 023508 (2006) [arXiv:hep-th/0605244].
11. M. S. Sloth, Nucl. Phys. B **748**, 149 (2006) [arXiv:astro-ph/0604488].
12. M. S. Sloth, Nucl. Phys. B **775**, 78 (2007) [arXiv:hep-th/0612138].
13. D. Seery, arXiv:0707.3377 [astro-ph].
14. D. Seery, arXiv:0707.3378 [astro-ph].
15. D. H. Lyth, arXiv:0707.0361 [astro-ph].
16. P. R. Jarnhus and M. S. Sloth, arXiv:0709.2708 [hep-th].
17. D. Wands, K. A. Malik, D. H. Lyth and A. R. Liddle, Phys. Rev. D **62**, 043527 (2000) [arXiv:astro-ph/0003278].
18. K. A. Malik and D. Wands, Class. Quant. Grav. **21**, L65 (2004) [arXiv:astro-ph/0307055].
19. D. H. Lyth, K. A. Malik and M. Sasaki, JCAP **0505**, 004 (2005) [arXiv:astro-ph/0411220].
20. B. L. Hu and E. Verdaguer, Living Rev. Rel. **7**, 3 (2004) [arXiv:gr-qc/0307032].

21. B. L. Hu, Physica A **158** (1989) 399.
22. B. L. Hu, Int. J. Theor. Phys. **38**, 2987 (1999) [arXiv:gr-qc/9902064].
23. R. Martin and E. Verdaguer, Phys. Rev. D **60**, 084008 (1999) [arXiv:gr-qc/9904021].
24. R. Martin and E. Verdaguer, Phys. Rev. D **61**, 124024 (2000) [arXiv:gr-qc/0001098].
25. B. L. Hu and E. Verdaguer, Class. Quant. Grav. **20**, R1 (2003) [arXiv:gr-qc/0211090].
26. B. L. Hu, A. Roura and E. Verdaguer, Phys. Rev. D **70**, 044002 (2004) [arXiv:gr-qc/0402029].
27. A. Roura and E. Verdaguer, Int. J. Theor. Phys. **39**, 1831 (2000) [arXiv:gr-qc/0005023].
28. A. Roura and E. Verdaguer, arXiv:0709.1940 [gr-qc].
29. J. S. Schwinger, J. Math. Phys. **2** (1961) 407.
30. K. c. Chou, Z. b. Su, B. l. Hao and L. Yu, Phys. Rept. **118** (1985) 1.
31. R. D. Jordan, Phys. Rev. D **33** (1986) 444.
32. E. Calzetta and B. L. Hu, "Cosmological Back Reaction Problems," Phys. Rev. D **35**, 495 (1987).
33. Z. b. Su, L. y. Chen, X. t. Yu and K. c. Chou, Phys. Rev. B **37**, 9810 (1987).
34. A. A. Starobinsky, *In *De Vega, H.j. (Ed.), Sanchez, N. (Ed.): Field Theory, Quantum Gravity and Strings*, 107-126*
35. V. F. Mukhanov, H. A. Feldman and R. H. Brandenberger, Phys. Rept. **215**, 203 (1992).
36. M. Sasaki, Prog. Theor. Phys. **76**, 1036 (1986).
37. H. Kodama and M. Sasaki, Prog. Theor. Phys. Suppl. **78** (1984) 1.
38. D. H. Lyth and D. Wands, Phys. Rev. D **68**, 103515 (2003) [arXiv:astro-ph/0306498].
39. Y. Urakawa and K. Maeda, [arXiv:0710.5342].
40. D. Polarski and A. A. Starobinsky, Class. Quant. Grav. **13**, 377 (1996) [arXiv:gr-qc/9504030].
41. C. Kiefer, I. Lohmar, D. Polarski and A. A. Starobinsky, Class. Quant. Grav. **24**, 1699 (2007) [arXiv:astro-ph/0610700].
42. R.Feynman and F.Vernon, Ann. Phys.(NY)**24**,118(1963), R. Feynman and A.Hibbs, Quantum Mechanics and Path Integrals (McGraw-Hill, New York, 1965), A. O. Caldeira and A. J. Leggett, Physica A **121**, 587 (1983), B. L. Hu and A. Matacz, Phys Rev. D **49**, 6612 (1994).

VARIATIONAL CALCULATION FOR THE EQUATION OF STATE OF NUCLEAR MATTER TOWARD SUPERNOVA SIMULATIONS

HIROAKI KANZAWA

Department of Pure and Applied Physics, Science and Engineering, Waseda University,
3-4-1 Okubo Shinjuku-ku, Tokyo 169-8555, Japan
Email: kann-sein@toki.waseda.jp

KAZUHIRO OYAMATSU

Department of Media Production and Theories, Aichi Shukutoku University,
Nagakute-cho, Aichi 480-1197, Japan
Email: oyak@asu.aasa.ac.jp

KOHSUKE SUMIYOSHI

Numazu College of Technology,
Ooka 3600, Numazu, Shizuoka 410-8501, Japan
Email: k-sumi@abox4.so-net.ne.jp

MASATOSHI TAKANO

Research Institute for Science and Engineering, Waseda University,
3-4-1 Okubo Shinjuku-ku, Tokyo 169-8555, Japan
Email: takanom@waseda.jp

An equation of state (EOS) for uniform nuclear matter is constructed at zero and finite temperatures with the variational method starting from the realistic nuclear Hamiltonian, and applied to the Thomas-Fermi (TF) calculations for atomic nuclei, toward a nuclear EOS for supernova simulations. For uniform nuclear matter at zero temperature, the two-body energy is evaluated in the two-body cluster approximation and the contribution of the three nucleon interaction (TNI) is taken into account phenomenologically. Parameters included in the TNI energy are determined so as to reproduce the empirical saturation conditions. At finite temperatures, a variational method by Schmidt and Pandharipande is employed to evaluate the free energy and related thermodynamic quantities. In the TF calculations for atomic nuclei, the parameters in the uniform EOS are tuned to reproduce empirical masses and RMS charge radii for β-stable nuclei.

1. Introduction

The necessity of researches on the equation of state (EOS) of nuclear matter is increasing in astrophysics. Especially, the nuclear EOS at finite temperatures plays important roles for studies of supernova (SN) explosions and related phenomena. However, the set of nuclear EOS's available for simulations of these phenomena is limited. An example is the one by Lattimer and Swesty[1] based on an extension of the compressible liquid drop model and another is by Shen et al.[2] based on the relativistic mean field theory. The latter is extended recently by taking into account the hyperon mixing[3] and hadron-quark phase transition.[4] Since these sets are obtained in phenomenological frameworks, the EOS based on the microscopic many-body approach starting from the realistic nuclear Hamiltonian is desirable.

In this situation, we undertake to construct a nuclear EOS for SNe using the variational method. In this paper, we report the present status of this undertaking. In Section 2, we construct the EOS for uniform nuclear matter at zero temperature, and extend the variational calculations to finite temperatures in Section 3.[5] Then, we apply the obtained EOS to isolated atomic nuclei using the Thomas-Fermi calculations, and tune parameters in the EOS to reproduce empirical data for β-stable nuclei in Section 4. This tuning of the parameters is important for the reliable SN-EOS, because SN matter is nonuniform at low densities. Concluding remarks are given in Section 5.

2. Uniform Matter at Zero Temperature

In this section, we calculate the energies for uniform matter at zero temperature. The nuclear Hamiltonian is expressed as follows:

$$H = H_2 + H_3. \tag{1}$$

Here, H_2 is the two-body Hamiltonian without the three-nucleon interaction (TNI):

$$H_2 = -\frac{\hbar^2}{2m}\sum_{i=1}^{N}\nabla_i^2 + \sum_{i<j}^{N} V_{ij}, \tag{2}$$

where m is the nucleon mass and V_{ij} is the two-body potential for which the isoscalar part of the AV18 potential is chosen. In Eq. (1), H_3 is the three-body Hamiltonian caused by TNI. In this study, the UIX potential is employed as TNI.

First, we calculate the expectation value of H_2 using the Jastrow-type wave function:

$$\Psi = \mathrm{Sym}\left[\prod_{i<j} f_{ij}\right]\Phi_{\mathrm{F}}. \tag{3}$$

Here, f_{ij} is the correlation function, and Φ_{F} is the Fermi-gas wave function at zero temperature. The correlation function f_{ij} is given by

$$f_{ij} = \sum_{t=0}^{1}\sum_{s=0}^{1}\left[f_{\mathrm{C}ts}(r_{ij}) + s f_{\mathrm{T}t}(r_{ij}) S_{\mathrm{T}} + s f_{\mathrm{LS}t}(r_{ij})(\boldsymbol{L}\cdot\boldsymbol{s})\right] P_{tsij}, \tag{4}$$

where $f_{\mathrm{C}ts}(r)$, $f_{\mathrm{T}t}(r)$ and $f_{\mathrm{LS}t}(r)$ are the spin-isospin-dependent central, tensor and spin-orbit correlation functions, respectively. S_{T} is the tensor operator, $(\boldsymbol{L}\cdot\boldsymbol{s})$ is the spin-orbit operator and P_{tsij} is the spin-isospin projection operator. We evaluate the expectation value of H_2 per nucleon in the two-body cluster approximation, i.e., E_2/N. Then, we minimize E_2/N with respect to variational functions, $f_{\mathrm{C}ts}(r)$, $f_{\mathrm{T}t}(r)$ and $f_{\mathrm{LS}t}(r)$, by solving the Euler-Lagrange equations.

In the minimization of E_2/N, we impose two constraints on the variational functions. The first is the extended Mayer's condition, which is a certain normalization condition. The second is the healing-distance condition, where the healing distance r_{h} is introduced such that $f_{\mathrm{C}ts}(r) = 1$ and $f_{\mathrm{T}t}(r) = f_{\mathrm{LS}t}(r) = 0$ at $r \geq r_{\mathrm{h}}$. In this study, it is assumed that $r_{\mathrm{h}} = a_{\mathrm{h}}$ $[3/(4\pi\rho)]^{1/3}$ with ρ being the nucleon number density, and the value of a_{h} is chosen so that the obtained E_2/N for symmetric nuclear matter is in good agreement with the results using a sophisticated variational calculation by Akmal, Pandharipande and Ravenhall (APR).[6] We note that it is difficult to extend the method by APR to finite temperatures. In this study, therefore, we employ a simpler variational method in order to treat nuclear matter at zero and finite temperatures in a consistent way.

Next, we consider the contribution of TNI, E_3/N, to obtain the total energy per nucleon $E/N = E_2/N + E_3/N$. In this study we decompose H_3 into the repulsive part H_3^{R} and the two-π-exchange part $H_3^{2\pi}$. Then, we express E_3/N as

$$\frac{E_3}{N} = \alpha\frac{\langle H_3^{\mathrm{R}}\rangle}{N} + \beta\frac{\langle H_3^{2\pi}\rangle}{N} + \gamma\rho^2 e^{-\delta\rho}. \tag{5}$$

Here, $<\quad>$ represents the expectation value using the Fermi-gas wave function. E_3/N includes four parameters α, β, γ and δ. In this study, we assume that α and β are common to symmetric nuclear matter and neutron matter, while $\gamma = 0$ for neutron matter. The values of these parameters are chosen so that E/N reproduces the empirical saturation density ρ_0= 0.16 fm^{-3}, saturation energy E_0/N = -15.8 MeV, incompressibility K = 250 MeV and symmetry energy $E_{\rm sym}/N$ = 30 MeV.

Figure 1 shows E_2/N and E/N for symmetric nuclear matter and neutron matter. The results by APR are also shown. It is seen that E/N for symmetric nuclear matter and neutron matter are in fair agreement with those by APR.

We apply the obtained nuclear EOS to neutron stars. Figure 2 shows the obtained neutron star mass as a function of the central mass density $\rho_{\rm m0}$. The maximum mass is about $2.22M_\odot$ ($M_\odot$ is the solar mass). We note that causality is violated at $\rho_{\rm m0} \geq \rho_{\rm mC} \sim 1.91 \times 10^{15}$g/cm^3. At the critical density $\rho_{\rm mC}$, the neutron star mass is $2.16M_\odot$.

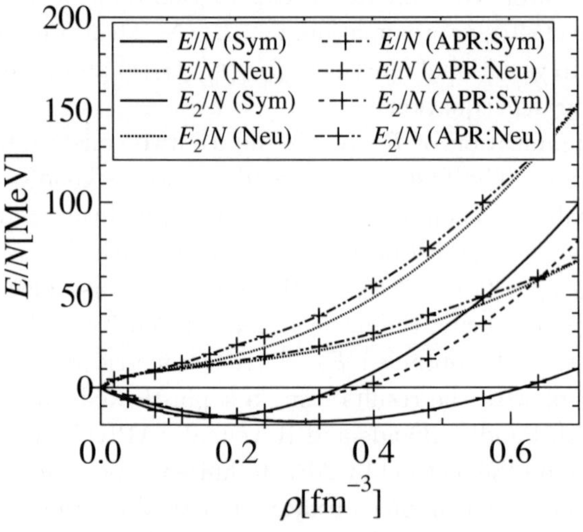

Fig. 1. Energies per nucleon without TNI, E_2/N, and with TNI, E/N, for symmetric nuclear matter and neutron matter as functions of the density. Corresponding energies obtained by APR are also shown.

3. Uniform Nuclear Matter at Finite Temperatures

In this section, we calculate the free energies per nucleon F/N for uniform nuclear matter at finite temperatures following the variational procedure proposed by Schmidt and Pandharipande.[7] In this method, F/N can be

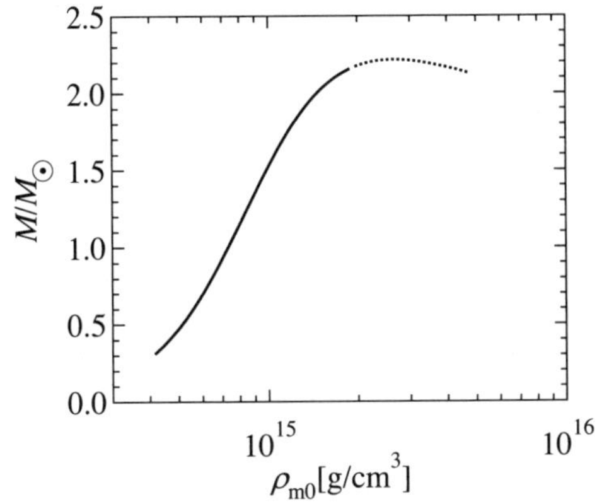

Fig. 2. Neutron star mass as a function of the central matter density. The dotted line shows the causality-violated solutions.

expressed as

$$\frac{F}{N} = \frac{E_{\mathrm{T0}}}{N} - T\frac{S_0}{N}. \tag{6}$$

Here, E_{T0}/N is the approximate internal energy and S_0/N is the approximate entropy at temperature T. E_{T0}/N is expressed as the sum of E_{T2}/N and E_3/N. E_{T2}/N is the two-body cluster approximation of the expectation value of $H_2$with the Jastrow-type wave function at finite temperatures:

$$\Psi = \mathrm{Sym}\left[\prod_{i<j} f_{ij}\right] \Phi_{\mathrm{F}}(n(k)). \tag{7}$$

For simplicity, f_{ij} in Eq. (7) is assumed to be the same as at zero temperature. $\Phi_{\mathrm{F}}(n(k))$ is the Fermi-gas wave function at finite temperature, which is specified by the averaged occupation probability of the quasi-nucleon states:

$$n(k) = \left(1 + \exp\left[\frac{(\epsilon(k) - \mu)}{k_{\mathrm{B}}T}\right]\right)^{-1}. \tag{8}$$

In Eq. (8), k is the wave number, k_{B} is the Boltzmann constant and $\varepsilon(k)$ is the quasi-nucleon energy:

$$\varepsilon(k) = \hbar^2 k^2/2m^*, \tag{9}$$

with m^* being the effective mass of the quasi-nucleon. In Eq. (8), μ is determined by the normalization condition on $n(k)$. The TNI contribution,

E_3/N, is assumed to be the same as at zero temperature, for simplicity. The approximate entropy S_0/N is

$$\frac{S_0}{N} = -\frac{k_\mathrm{B}}{N} \sum_i [\{1 - n(k)\} \ln \{1 - n(k)\} + n(k) \ln n(k)]. \tag{10}$$

Then, we minimize F/N with respect to m^*.

The obtained F/N is shown in Fig. 3. Also shown in Fig. 3 is F/N by Friedman and Pandharipande (FP).[8] At low densities, our result is in good agreement with that by FP, while our F/N is higher than that by FP at high densities, which implies that our EOS is stiffer than that by FP. Figures 4 shows the entropy S/N; the approximate entropy S_0/N is also shown. It is seen that S/N is very close to S_0/N, which implies that the variational calculation is self-consistent.

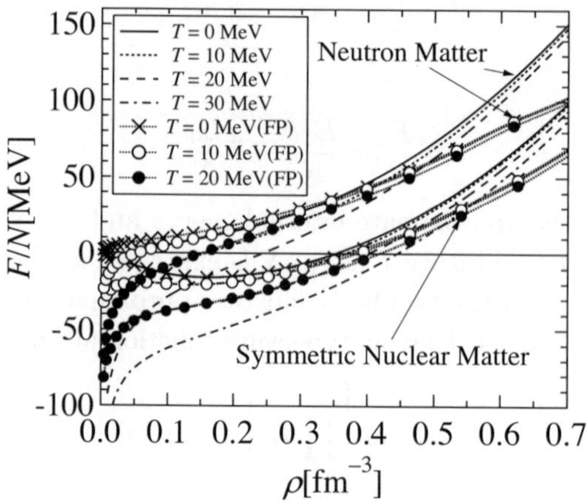

Fig. 3. Free Energies per nucleon, F/N, at T = 0, 10, 20 and 30 MeV for symmetric matter and neutron matter as functions of the density. The free energies obtained by FP are also shown.

4. Thomas-Fermi Calculations for Atomic Nuclei

In the SN explosions, as mentioned in Section 1, the density distribution of nuclear matter is nonuniform at low densities. We are planning to treat nonuniform nuclear matter using a simplified Thomas-Fermi (TF) approximation with the EOS for uniform nuclear matter obtained above. Before treating nonuniform nuclear matter, in this section, we perform the TF calculations for atomic nuclei, and tune the parameters in the EOS to repro-

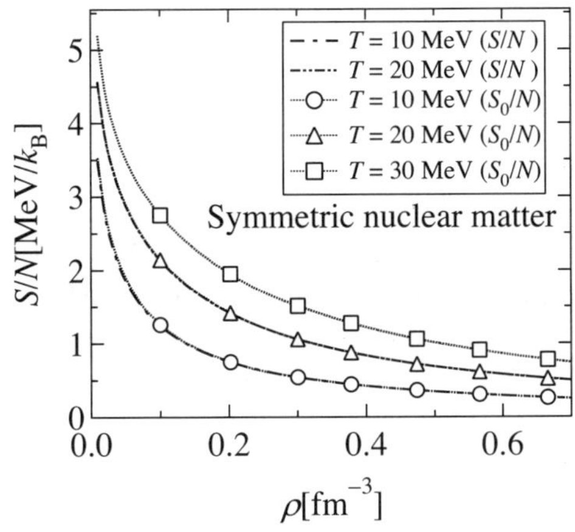

Fig. 4. Entropies per nucleon, S/N, at $T = 10$ and 20 MeV for symmetric matter as functions of the density. The approximate entropies per nucleon S_0/N are also shown.

duce empirical data of nuclear masses, RMS charge radii and the position of the β-stability line: The EOS reproducing those empirical data for known nuclei will be more reliable.

In this study, we employ the TF approximation used by Oyamatsu[9] and express the binding energy $B(N, Z)$ of a nucleus with the proton number Z and the neutron number N as

$$-B(N,Z) = \int dr\varepsilon(\rho_{\rm n}(r), \rho_{\rm p}(r)) + F_0 \int dr \left|\nabla\rho(r)\right|^2 + \frac{e^2}{2}\int dr \int dr' \frac{\rho_{\rm p}(r)\rho_{\rm p}(r')}{|r-r'|}. \quad (11)$$

The first term on the right-hand side of Eq. (11) is the gross energy, and $\varepsilon(\rho_{\rm n}, \rho_{\rm p})$ is the energy density of uniform nuclear matter for neutron number density $\rho_{\rm n}$ and proton number density $\rho_{\rm p}$. $\varepsilon(\rho_{\rm n}, \rho_{\rm p})$ is constructed using E/N obtained in Section 2. The second term represents the gradient energy with $\rho(r) = \rho_{\rm p}(r) + \rho_{\rm n}(r)$, and $F_0 = 70$ MeVfm^5. The third term represents the Coulomb energy. Here, we assume that the nucleus is spherical and that $\rho_i(r)$ $(i = {\rm p, n})$ is expressed as follows:

$$\rho_i(r) = \begin{cases} \rho_i^{\rm in}\left[1 - \left(\frac{r}{R_i}\right)^{t_i}\right]^3 & (0 \le r \le R_i), \\ 0 & (R_i \le r). \end{cases} \quad (12)$$

Then, we minimize $-B(N,\ Z)$ for each nucleus $(N,\ Z)$ with respect to the parameters $\rho_i^{\rm in}$, R_i and t_i (i = p, n) under the normalization conditions,

$$N = \int dr \rho_{\rm n}(r), \quad Z = \int dr \rho_{\rm p}(r). \tag{13}$$

In Figs. 5 and 6, the obtained masses and RMS charge radii for β-stable nuclei are compared with the smoothed empirical data given in Ref. 9. Figure 5 shows the deviation of the mass, $\Delta M = M_{\rm TF} - M_{\rm emp}$, (filled circles) as a function of the mass number A, where $M_{\rm TF}$ and $M_{\rm emp}$ are the calculated and smoothed empirical masses, respectively. It is seen that ΔM is about 6-18 MeV. The deviation of the RMS charge radii, $\Delta r_{\rm RMS} = r_{\rm TF} - r_{\rm emp}$, is shown in Fig. 6 (filled circles), where $r_{\rm TF}$ and $r_{\rm emp}$ are the calculated and empirical values, respectively. It is seen that $\Delta r_{\rm RMS}$ is about 0.06-0.08 fm.

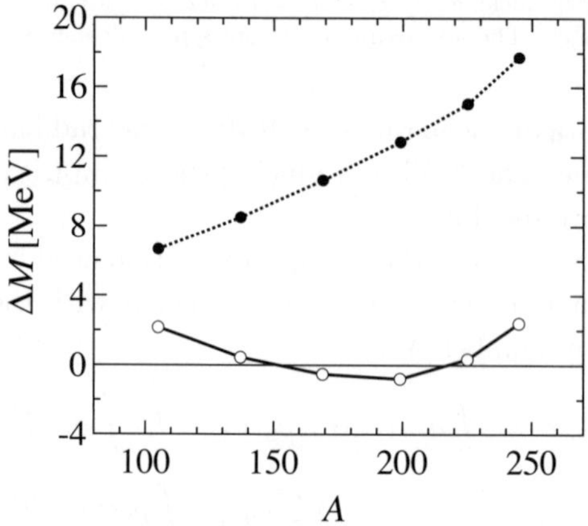

Fig. 5. Deviations of the masses from the empirical values, ΔM, as functions of the mass number. Filled circles are for the EOS in Section 2 and open circles are for the parameter-tuned EOS.

In order to reduce $|\Delta M|$ and $|\Delta r_{\rm RMS}|$, we tune the values of the parameters γ and δ in Eq. (5). ΔM and $\Delta r_{\rm RMS}$ after tuning the parameters are shown in Figs. 5 and 6 by open circles. We obtained $|\Delta M| < 3$MeV and $|\Delta r_{\rm RMS}| < 0.01$ fm. The tuning of γ and δ corresponds to the tuning of ρ_0 and E_0/N for symmetric nuclear matter: $\rho_0 = 0.165$ fm^{-3} and $E_0/N = -16.05$ MeV after tuning.

Using this tuned EOS, we consider the β-stability line. Figure 7 shows the calculated binding energies per nucleon for various nuclei. The empirical β-stability line is also shown. It is clearly seen that the calculated β-stability

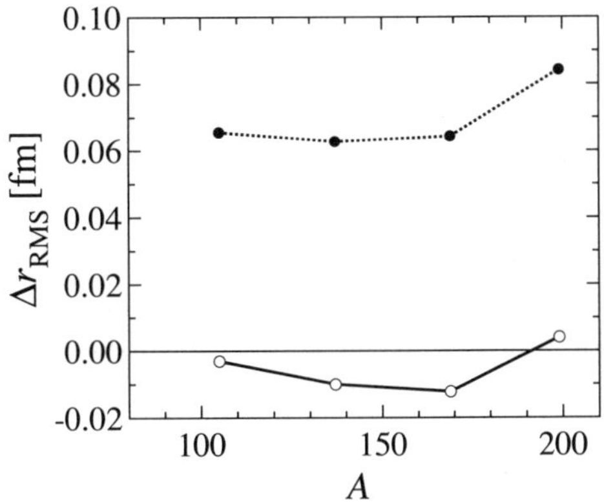

Fig. 6. Deviations of the RMS charge radii from the empirical values, $\Delta r_{\rm RMS}$, as functions of the mass number. Filled circles are for the EOS in Section 2 and open circles are for the parameter-tuned EOS.

line is on the neutron-rich side of the empirical one. This implies that the symmetry energy $E_{\rm sym}/N$ of our nuclear EOS is somewhat too low. It is also found that the position of the calculated β-stability line is almost insensitive to the tuning of γ and δ. In order to improve the β-stability line, tuning of the parameters including α and β in Eq. (5) is necessary.

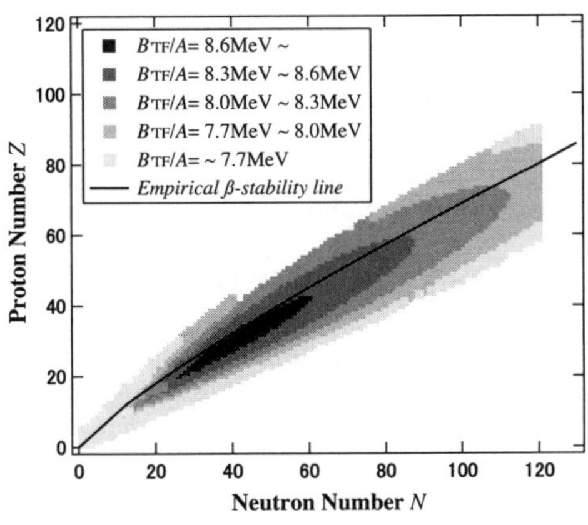

Fig. 7. Binding energies per nucleon. The empirical β-stability line is also shown.

5. Concluding Remarks

In this paper, we constructed an EOS for uniform nuclear matter at zero and finite temperatures starting from the realistic Hamiltonian, and tuned the parameters in the EOS to reproduce the empirical data for β-stable nuclei. The deviations of the masses and RMS charge radii decrease largely by tuning γgndgδ. We are planning to tune the parameters including α and β, in order to improve the β-stability line. Then, we proceed to the TF calculations for nonuniform nuclear matter, toward a nuclear EOS table for SN simulations.

Acknowledgments

This study is supported by a Grant-in-Aid for the 21st century COE program "Holistic Research and Education Center for Physics of Self-organizing Systems" at Waseda University, and Grants-in-Aid from the Scientific Research Fund of the JSPS (Nos. 18540291, 18540295 and 19540252). Some of the numerical calculations were performed with SR8000/MPP and SR11000/J2 at the Information Technology Center of the University of Tokyo.

References

1. J. M. Lattimer and F. D. Swesty, *Nucl. Phys. A* **535**, 331 (1991).
2. H. Shen, H. Toki, K. Oyamatsu and K. Sumiyoshi, *Prog. Theor. Phys.* **100**, 1013 (1998).
3. C. Ishizuka, A. Ohnishi, K. Sumiyoshi and S. Yamada, *Prog. Theor. Phys. Suppl.* **156**, 152 (2004).
4. K. Nakazato, K. Sumiyoshi and S. Yamada, submitted to *Phys. Rev. D*.
5. H. Kanzawa, K. Oyamatsu, K. Sumiyoshi and M. Takano, *Nucl. Phys. A* **791**, 232 (2007).
6. A. Akmal, V. R. Pandharipande and D. G. Ravenhal, *Phys. Rev. C***58**, 1804 (1998).
7. K. E. Schmidt and V. R. Pandharipande, *Phys. Lett.* **87 B**, 11 (1979).
8. B. Friedman and V. R. Pandharipande, *Nucl. Phys. A* **361**, 502 (1981).
9. K. Oyamatsu, *Nucl. Phys. A* **561**, 431 (1993).

TWO STRONG RADIO BURSTS AT HIGH AND MEDIUM GALACTIC LATITUDE

SUMIKO KIDA

Astrophysics, Department of Pure and Applied Physics Graduate School of Science and Engineering, Waseda University, 3-4-1 Okubo, Shinjuku-ku, Tokyo, 169-8555, Japan
E-mail: sumiko_k@suou.waseda.jp
http://www.astro.phys.waseda.ac.jp

TSUNEAKI DAISHIDO

Department of Science School of Education, Waseda University, 1-6-1, Nishi-Waseda, Shinjuku-ku, Tokyo, 169-8050, Japan

We constructed eight 20m elements in Nasu, and are observing radio transients over a wide-field at 1400 MHz. We report on two radio transients detected on consecutive drift scanning observations at declination 32 degrees over about two months. One of the two transients "WJN J1039+3200" appeared at $\alpha=10^h39^m40^s \pm 10^s, \delta = 32° \pm 0.4°$ on March 4, 2005 and the other one "WJN J0645+3200" appeared at$\alpha=06^h45^m25^s \pm 10^s, \delta = 32° \pm 0.4°$ on March 24, 2005DBoth exhibited flux densities in excess of 1 Jy. The burst duration was within two days. There are few examples of radio transients outside the Galactic plane. Therefore, these are very important observations. We have already reported on four radio transients with features that looks like the two transients detected this time. The duration of five transients is within two days, and one transient is within three days. Four transients of six transients were detected at a high galactic latitude of b>30°. Counterparts of six WJN Transients were included X-ray sources in four events, and had a consistency of 66%. The consistency of γ-ray, PGC Galaxy, NVSS and FIRST source was concentrated at about 50%. We were not able to find a special feature in the counterpart The distribution was verified by making a logN-logS plot using those data and the newly detected data. As a result, the distribution of the radio transients that we observed might have an isotropic distribution not dependent on Galactic longitude and Galactic latitude. We cannot yet clarify these two radio transients, because their features are different from any radio burst observed in the past.

Keywords: Stars:variables:other; radio coutinum:general.

1. Introduction

A lot of transient phenomena have been observed by UHURU, Newton, ROSAT and INTEGRAL, etc. since the beginning of the 1970's. However, almost all observations have been transient X-rays and γ-rays. There have been very few radio observations in the past. A large-scale radio burst Cyg X-3 was observed by Daishido et al. (1983) and Gregory (1972, 1975). In addition, radio observations were made at Green Bank Telescope by Gregory and Taylor (1981, 1986), and VLA by Hyman etc. (2005). These observations detected the burst of LSI+61 303, the radio transient GCRT J1745-3009 of 1 Jy class and so on. The archival study of NVSS and FIRST by Levinson et al. (2002) and follow-up by Gal-Yam et al. (2006) who identified several radio transients. Recently, Rotating Radio Transients (RRATs) were detected by the radio telescope of Parkes (Mclaughlin et al. 2006). Ten radio transients were detected from archival data of VLA (Bower et al. 2007). The origin of these ten radio transients is unknown.

The radio bursts detected in the past were observed mostly at the Galactic center and Galactic plane. Observation of these regions has been emphasized, while the high Galactic region has been neglected. With the interferometer that we constructed in NASU, the high Galactic region can be observed in a manner similarly to observations in the Galactic plane (Daishido et al. 1996, 2000; Ichikawa et al. 2004, Takeuchi et al. 2005).

Sec. 2 describes the outline of WASEDA NASU Pulsar Observatory, and Sec. 3 describes the method of analyzing data. Finally, our conclusions concerning the two radio transients and four transients that has already been reported are described in Secs. 4 and 5.

2. Observation

WASEDA NASU Pulsar Observatory was originally constructed by our laboratory. Therefore, observations appropriate for our research can be carried out at any time. In WASEDA Observatory, eight 20 m-diameter spherical dish antennas are lined from east to west, and can observe continuously for 24 hours as two elements $\times$ 4 interferometers. Four declinations can be observed at the same time. Our observation method is the drift-scanning by the phase switching method. The observation frequency is 1400 MHz and the bandwidth is 20 MHz. The observable declination is $32° < \delta < 42°$.

This report concerns two radio transients detected by continuous observation at declination 32° from Feb to Apr, 2005.

3. Analysis

When the electromagnetic radiation from the object on the celestial sphere is detected in the interferometer, the output pattern is in the shape of waves of stripes that appears as the time change is obtained. The shape of waves of stripes pattern (fringe) is analyzed. The fringe cycle that comes from the object on the celestial sphere is decided by the baseline and declination. The baseline is 84 m in WASEDA observatory, and declination is 32 in this observation. The fringe cycle calculated from this condition, is approximately 40 sec. Therefore, if it is a signal from the object on the celestial sphere, it is sure to be approximately 40 sec.

First, data is processed with the Automatic Burst Search Software (Kuniyoshi et al. 2006) over one day. When processing with this software, it is easy to search for a transient fringe because the fringe is plainly visible. Next, the autocorrelation $\int_0^{200}(f(t)f(t-T))dt$ (200 s within the range of integration was decided by HPBW.) in the fringe is taken, and the fringe cycle of the standard source DA344 (4631 mJy) is compared with the fringe cycle of the radio transient(Fig.1). As a result of the autocorrelation, the fringe cycle of the standard source DA344 is approximately 40 sec. On the other hand, this correlation is not seen from the coordinate data of WJN J1039+3200 and WJN J0645+3200 because of the noise on most days. However, a fringe of 40 sec similar to DA 344 was observed on March 4, 2005 for WJN J1039+3200, and on March 24, 2005 for WJN 0645+3200 (Fig. 1). It is clear that these two outbreak fringes have their origin in the radio transient source. In addition, not only radio but also X-rays and γ-rays verify the counter part in the error box of the radio transient.

It is not possible that this fringe is not the one on the celestial sphere but due to other objects (e.g. airplane, satellite, and other artificial things). It is not possible because a steady fringe could not last for 200 seconds because of the drift scanning method of observation. If it is a planetary exploration satellite in the distance, the signal would be detected every day in similar coordinates without fail, because you can hardly shift position. Moreover, if it is a fixed satellite, the signal would always be detected. If it is due to orbiting satellites, a constant fringe of 200 seconds would not be detected because the speeds are faster than the rotations of the celestial sphere and the fringe cycle would become shorter. For a more interference, a strong signal enters all antennas under the observation, and they would not be observed as a constant fringe.

The error box of the data of the WASEDA observatory is described as follows. The error in the direction of R.A. is approximately 10 sec. This

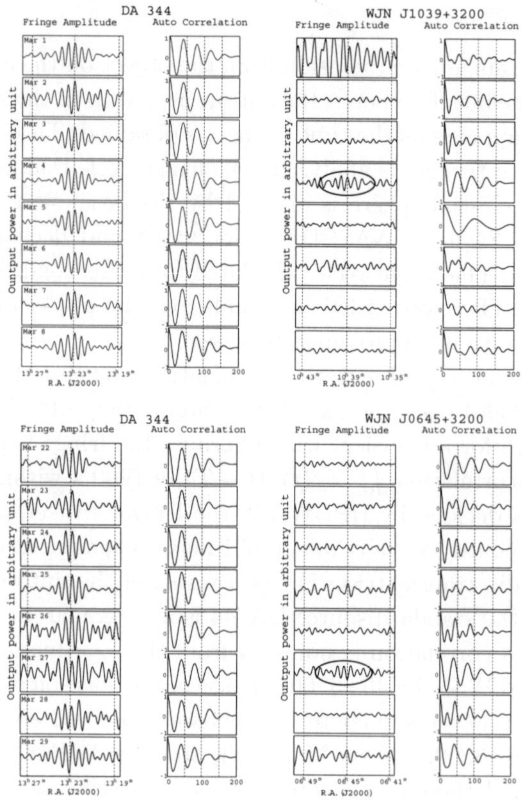

Fig. 1. Above: Daily fringe data of radio transient WJN J1039+3200 and that of a standard source DA 344. WJN J1039+3200 appeared on March 4. Below: Daily fringe data of radio transient WJN J0645+3200 and that of a standard source DA 344. WJN J0645+3200 appeared on March 27. In the fringe amplitude data x-axis is right ascension and y-axis is output power in arbitarary units. In the autocorrelation data,the x-axis is dispalacement [s] and the y-axis is dimensionless.

depends on the signal to noise ratio. The error in the declination was limited to 0.8° by HPBW. We searched for counterparts within the error box in the catalog and database (NED, HEASARC, etc) of not only radio but also$X$-rays, γ-rays and Galaxy. The counterpart is evaluated in Sec. 5

4. Results

We detected two radio transients (WJN J1039+3200 and WJN J0645+3200), with declination 32 degrees from February to April, 2005. As a result of the data analysis, it turned out that the two radio transients had the same fringe cycle as standard source DA344 (Fig. 1). Table 1 shows the profile of the two radio transients. The flux density was determined according to the power spectrum ratio of the transient source to the standard source. The burst duration had a range of from four minutes to two days. The transients can be compressed only to this range for the drift scanning observation in order to give priority to the wide-field observation. As previously described, four minutes is the time required for the source to pass HPBW.

For counterparts, PGC Galaxy, the bright radio source, the faint radio source are within WJN J1039+3200 positional error and PGC Galaxy, X-ray source, the bright radio source are within WJN J0645+3200 positional error. In the following 5.Discussion, these counterparts are discussed in detail with four WJN Radio Transients that has already been made public.

Moreover, neither WJN J1039+3200 nor WJN J0645+3200 were detected in the follow-up observation that was carried out from 10 to 26 November, 2006.

Table 1. Profile for WJN Radio Transients.

Profile for WJN J1039+3200	
Equatorial Coordinate (J2000)	α=10h39m43s±10s , δ=32°00'00"±0.4°
Galactic Coordinate	l=194°39', b=+66° 12'
DATE	2005/03/04 UT13:43:20
The Burst Duration	4[min]~2[day]
The Flux	~1700[mJy] (21[cm])
Profile for WJN J0645+3200	
Equatorial Coordinate (J2000)	α=06h45m15s±10s , δ=32°00'00"±0.4°
Galactic Coordinate	l=183°18', b=+12°47'
DATE	2005/03/27 UT 08:34:40
The Burst Duration	4[min]~2[day]
The Flux	~1200[mJy] (21[cm])

Note: Positions(J2000),and flux density is 1.4 GHz. The error range of coordinates of the direction of RA is decided by Signal-Noise ratio, and the direction of Dec is decided by HPBW.

5. Discussion and Conclusion

5.1. *New type radio transients*

We compared the radio transients with features of various radio transients that have been reported in the past in order to clarify the origin of the WJN radio transients that we detected. We thought that possibly the WJN radio transients were from a flare star, Rotating Radio Transients (RRATs) and

GRB radio afterglows. Those possibilities are discussed below.

First, we assumed that WJN radio transients originated from a flare star or a brown dwarf. AD Leo (Jackson et al. 1989), L726-8A (Stephen et al. 1986), and LP944-20 (Beger et al. 2001) were observed in the vicinity of 1400 MHz. According to those reports, the flux density was in the several mJy range, and the burst duration was from minutes to hours. Because a flare star and a brown dwarf in the vicinity of a solar system radiate chiefly from tens of MHz to hundreds of MHz, the flux density should be very weak at our observational frequency of 1400 MHz. The flux densities of WJN radio transients exceed 1 Jy, therefore flare stars and brown dwarfs can not be the origins of this radiation.

Second, we considered the possibility of RRATs discovered by the Parkes pulsar survey at 1400 MHz (McLaughlin et al. 2006). The flux density of RRATs was 0.1-3.6 Jy at 1400 MHz. Because a strong electric wave is radiated from the rotating neutron star producing the RRATs, it is periodically observed. The periodicity is about 0.4-7 seconds, and the burst duration is 2-30 milliseconds. Moreover, it has been confirmed that RRATs are concentrated in the Galactic plane. The duration time of RRATs is much shorter than WJN radio transients, and distribution is concentrated in the Galactic plane. Therefore, RRATs can not be the origin of this radiation.

Third, we considered the possibility of GRB radio afterglow because WJN radio transients were observed at medium and high Galactic latitude. However the radio afterglows are continuously observed from 1 to 3 months, and the flux densities were from micro-Jy to milli-Jy, according to a past observation. WJN radio transients will be different from these.

In addition, we assumed that WJN radio transients were due to the influence of the scintillation and the effect of the gravitational lens. Though the change of the flux density by scintillation was reported by Rickett et al. (1995) and Jauncey et al. (2001), the variation at 1400 MHz is at most about 10%. This does not correspond to a big variation of WJN radio transients. If, depending on the effect of the gravitational lens, it doesn't depend on the observational frequency, then the same variation should be observed. And the light curve is sure to become symmetry for the eclipse. Because we have carried out drift scanning observations, a detailed light curve cannot be obtained. But it is likely to be observed not as a burst only during a day like WJN radio transients but as a slower change. Therefore, these possibilities are also untenable.

We make the assumption from the above discussion that the radio source for the WJN radio transients has as yet not been clarified. As described in Sec.1, a lot of radio observations at Galactic center and Galactic plane have been performed. However, wide-field observation of the area left of the Galactic plane has received little attention. Therefore, the radio transient at high and medium Galactic latitude like WJN radio transients might not have been observed. Our observation is very rare and important. Because WJN J1039+3200 was located at high Galactic latitude, we consider the possibility that an active source near the solar system or the usually quiet extragalactic source is related. WJN J0645+3200 was located at medium Galactic latitude, and it was located in the vicinity of the Gould Belt. There might be a possibility of radio transients related to WJN because the star formation is an active area in the Gould Belt. Anyway, it is still too early for a decision on the origin. We want to clarify the origin carefully with future observations.

5.2. *New population*

Several radio transients with features similar to WJN J1039+3200 and WJN J0645+3200 which are reported here have been detected in the Waseda Nasu observation (Kuniyoshi et al. 2007; Matsumura et al. 2007; Niinuma et al 2007). Table 2 shows the profile of two radio transients that newly detected and four radio transients that have already been reported.

We synthesized these detection results, and performed statistical analysis. Isotropic distribution of WJN radio transients were verified by making a logN -logS plot. If sources are isotropically distributed, there is the characteristic that the slope is -3/2 when the logarithm of N (number of sources) is taken as the vertical axis and the logarithm of S (flux density in source) is taken as the horizontal axis. The volume where the source is included is proportional to the third power of the distance and the flux density in the source is proportional to the negative second power of the distance. Figure 2 shows a logN-logS plot of the WJN radio transients detected at WASEDA. The correlation coefficient r of the observed data and the slope of -3/2 is 0.93 in Fig.2. The correlation is higher than a logN-logS plot for GRB which is known as an isotropic distribution. As a result, it is quite reasonable to assume that WJN radio transients are isotropically distributed.

We keep the isotropically distribution discussion in mind and describe about the counterpart included in the error box We calculated the consistency of the counterpart of six WJN transients (Only the agreement or the disagreement was considered. The number of counterparts was not considered. For instance, even if one transient contained two X-ray sources, it was counted as one.) . 66% of the transients have error boxes consistent

Table 2. WJN Radio Transients.

Name	RA	Dec	b	Date	Flux[Jy]	Duration	Counterpart
WJN J0445+4130	$04^h45^m17^s$	$+41°30'$	$-2°$	2005/01/10	1.8	4m~2days	$X, \gamma \times 4$
WJN J0645+3200⋆	$06^h45^m15^s$	$+32°00'$	$+12°$	2005/03/27	1.2	4m~2days	Br, X, G
WJN J1039+3200⋆	$10^h39^m43^s$	$+32°00'$	$+66°$	2005/03/04	1.7	4m~2days	$Fr \times 2, Br \times 3, G$
WJN J1043+4130	$10^h43^m06^s$	$+41°00'$	$+60°$	2005/01/02	1.7	4m~2days	$\gamma \times 2$
WJN J1443+3439	$14^h43^m22^s$	$+34°39'$	$+65°$	2005/02/13,14	1.7,3.2	2days~3days	$Fr \times 4, Br \times 2, IR, X \times 2, \gamma \times 2$
WJN J1737+3808	$17^h37^m17^s$	$+38°08'$	$+30°$	2004/03/20	1	4m~2days	$Br, X, G \times 2$

Note: This table is a profile of six WJN transients. × in the counterpart column shows the number. ⋆ is indicates new transients presented in this report. Because WJN J1443+3539 was detected in two successive days, two values of flux are written.

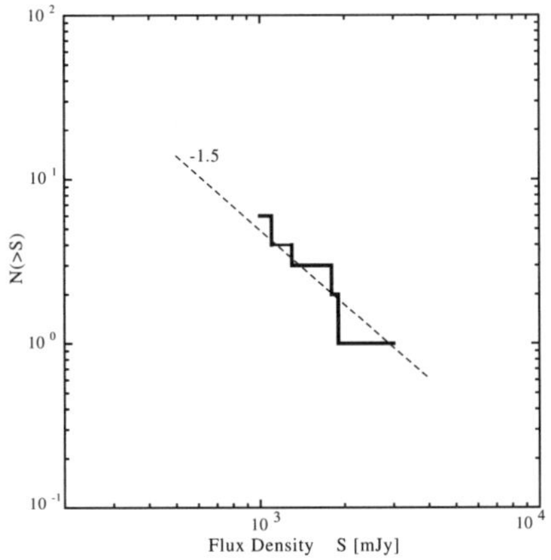

Fig. 2. logN-logS of WJN Radio Transient. This figure shows a high correlation between WJN Radio Transients and a straight line of slope -3/2 that shows isotropic distribution.

with an X-ray source, 50% are consistent with a γ-ray source, 50% are consistent with PGC galaxies, 50% are consistent with NVSS catalog sources and 33% are consistent with FIRST catalog sources. A remarkable feature is not seen though the consistency of the candidate of X-rays is the highest. Everything has been stored in about 50%.

In addition, we calculated how many counterparts for each transient (The number of counterparts was considered. If one transient contained to two X-ray sources, it was counted as two.). WJN transient has 0.8 X-ray sources per one event, 1.3 γ-ray sources per one event, 0.7 PGC Galaxies per one event, 1 NVSS source per one event, 1 FIRST source per one event. WJN transient has 5.1 counterparts per one event as a whole. For the comparison, from the density in the source recorded in these catalogs, we calculated an approximate probability that the source of these catalogs was detected in the error box in WASEDA. The error box in WASEDA is described with 3.Analays. We calculated the detection probability from the ratio of all sources of catalog and size of error box in WASEDA [0.0018deg^2]. The detection probability of PGC Galaxy are about 0.5 sources per one WASEDA error box, the detection probability of NVSS is about 1.1 sources, the detection probability of FIRST is about 1.6 sources. The detection probability of these catalogs and counterpart per one event were compared. It is almost corresponding. Therefore, we were not able to find a special feature of the counterpart.

The observable area in WASEDA observation is 5% of all celestial spheres. It is assumed that it is an isotropic distribution in all celestial spheres based on the observation that it is an isotropic distribution within our observable area. Then we could expect to observe approximately 100 radio transients like WJN sources per year. This it is important for us to obtain more samples of radio transients.

The feature in six Transients that should be emphasized is the observation on only day or two days Only one transient was detected in two successive days, the flux density was the strongest exceeding three Jy. The flux densities of other Transients during a day were about 1-1.8 Jy. We presume from the above-mentioned that the correlation of duration scale and burst strength of WJN Transients might tend toward strength increases through long duration.

There has not yet been further detection through follow-up observation. The clarification of the origin will advance if as much as only one burst is detected by follow-up observation, just as for the initial GRB observation. Since our intention is the clarification of WJN radio transients, we will keep wide-field observing at the radio wavelength in the future. The observation of radio transients like WJN will become more active by shifting in LOFAR and SKA age in the future. We have great expectations for surveys in these new ages.

This research received the support of the 21st century COE program in Waseda University. We wish to express our gratitude for this support.

References

1. Berger, E., et al. 2001, Nature, 410, 338
2. Bower, Geoffrey C., et al. 2007, APJ, 2007arXiv0705.3158B
3. Cordes, James M., Lazio, T, JosephW., McLaughlin, M, A., 2004, New AR, 48, 1459
4. Gal-Yam, A., et al. 2006, AJ, 639, 331
5. Hyman, S, D., Lazio, T, J, W., Kassim, N, E., & Bartleson, A. L., 2002, AJ, 123, 1497
6. Hyman, S, D., Lazio, T, J, W., Kassim, N, E., Ray, P. S., Markwardt, C. B., & Yusefzadeh, F., 2005, Nature, 434, 50
7. Kraus, J. D., 1986, Radio Astronomy 2nd edition (Cygnus-Quasar Books)
8. Kuniyoshi, M., et al. 2006, PASP, 118, 901
9. Matsumura, N., et al. 2007, AJ, 133, 1442
10. Levinson, A., et al. 2002, AJ, 576, 923
11. McLaughlin M, A., et al. 2006, Nature, 439 ,817
12. Niinuma, K., et al. 2007, APJ, 657, L37
13. Stephen, M. White., R, Kundu., Peter D, Jackson., 1986, APJ, 311 , 814
14. Takefuji, K., et al. 2007,PASP, 119, 1145

EFFECTS OF QCD PHASE TRANSITION ON THE EJECTED ELEMENTS FROM THE ENVELOPES OF COMPACT STARS

N. YASUTAKE* and S. YAMADA

Science and Engineering, Waseda University, Tokyo 169-8555, Japan
**E-mail: yasutake@heap.phys.waseda.ac.jp*
www.heap.phys.waseda.ac.jp/yasutake/opening.html

T. NODA and M. HASHIMOTO

Department of Physics, Kyushu University, Fukuoka 810-8560, Japan
E-mail: tsune@gemini.rc.kyushu-u.ac.jp

K. KOTAKE

Division of Theoretical Astronomy, National Astronomical Observatory of Japan, Tokyo 181-8588, Japan
E-mail: kkotake@yso.mtk.nao.ac.jp

We perform hydrodynamical collapse and bounce simulations of neutron stars induced by QCD phase transition. We study how the phase transition affects the elements of the envelopes and crusts of neutron stars near the epoch of core-bounce. We assume the first-order phase transition, a chirally symmetric quark-gluon plasma for quark phase. The initial elements are adopted from the realistic evolutional calculations of neutron stars containing the cooling effects and the nucleosynthesis of envelope such as type I X-ray burst. We find that the elements of envelope can eject directly (without nucleosynthesis) for such a collapse and bounce by the transition.

Keywords: QCD phase transition; nucleosynthesis; quark star; neutron star.

1. Introduction

It has been presented that the quark matter might appear during supernova explosions or the transition of a neutron star to a quark star. Some models to identify quark stars are suggested theoretically, such as mass–radius relation (e.g. for RXJ1856-37[1–4]), cooling process,[5–8] gravitational wave,[9,10] the maximum energy release estimations from neutron stars to quark stars[11,12] etc.. However, it is difficult to distinguish quark stars and neutron stars clearly.

On the one hand, there has been extensive works devoted to studying evolution of compact stars in the context of cooling processes and X-ray bursts.[13,14] As for X-ray bursts, it has been widely accepted that type I X-ray bursts from low mass X-ray binaries (LMXBs) are due to thermonuclear runaways in accreted materials on the surface of compact stars.[15,16] Koike et al.[14] and Woosley et al.[17] has suggested that the envelopes of compact stars might be the sites to produce p-process or rp-process elements. However, there have been few studies about the ejection mechanisms of such elements, since the most mass of neutron stars falls to black holes after the collapses[18]

In this paper, we use hydrodynamics code, and estimate the possibility of the ejection of such elements from compact stars with QCD phase transition.

2. Input Physics

Our initial model is a realistic spherical neutron star including the cooling effects and/or the nucleosynthesis of envelope such as X-ray burst.[14] The composition is displayed in Fig. 1. This composition includes characteris-

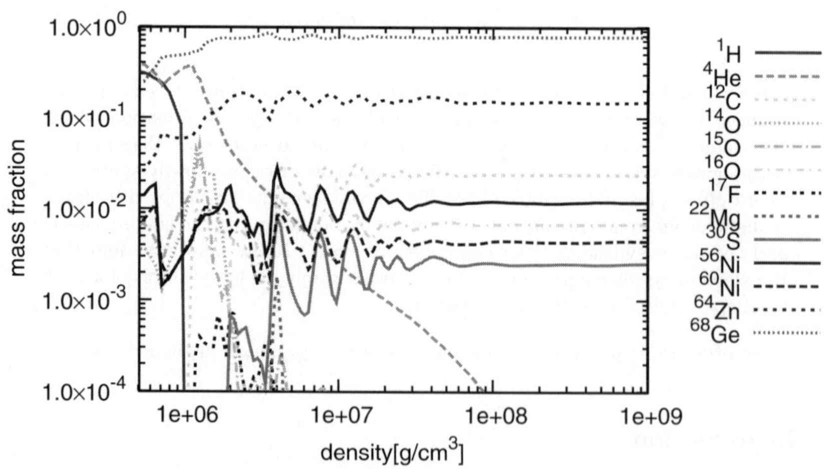

Fig. 1. Mass fractions of the envelope elements on initial neutron stars.

tic rp-process elements, such as ^{68}Ge, ^{64}Zn. There are very few simulations about the dynamical mass ejection of these elements from compact stars. We assume the uniform initial electron fraction, $Ye = 0.1$, and the uniform entropy distribution, $s_B = 0.01 k_B^{-1}$. The temperature at the envelope is about $\sim 10^5$ K, for this small entropy. Therefore, we assume the cold neutron stars as initial models.

Furthermore, we adopt the realistic equation of state (EOS) for the hadron matter.[19] This EOS is based on relativistic mean field theory. Since there does not exist realistic EOS that describes QCD phase transition, we use the simple MIT bag model. For mixed phase, we follow the general Gibbs condition considering electrons chemical potential appropriately, since the effect of the the general Gibbs condition was not trivial for EOSs.[20] We set the effective mass of u, d, and s-quarks as 5, 5, and 150 MeV, respectively. The bag constant set is $B = 150$ MeV/fm^3. This EOS is displayed in figure 2.

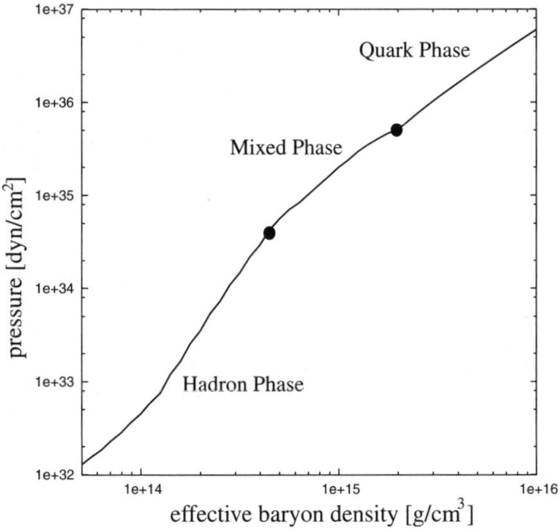

Fig. 2. Equation of state under the general Gibbs condition at $B = 150$ MeV/fm^3, $T = 0$ MeV. Mixed phase is between two dots ($\bullet$) which mean critical density points of QCD phase transition.

The numerical method in this paper is based on the ZEUS-2D code.[21] This code is two dimensional magneto-hydrodynamics code. However, we perform spherical symmetric test calculation in this paper. We take into account the general relativistic correction[22] to the Newtonian gravity. This correction is be considered reasonable and proper for spherical symmetric simulation. Though our purpose is the estimation of the ejection from the envelope, the density and other physical quantities are change drastically in the envelope of compact stars. Therefore, we adopt the nonuniform grid (see Fig. 3). The smallest size of the grids are about $\sim$ 1 meter. We set the such grids near the envelopes of compact stars. The largest size of the grids is about $\sim$ 100 meter. We set the such large grids near the core of the compact stars or the outside of the stars. This initial structure of the neutron star is perfectly stable during our calculational time scale without QCD phase transition (see the time of Fig. 4 (a)).

We do not consider neutrino effects, since the time scale of the collapse and bounce is very short, $\sim$ 0.1 ms.

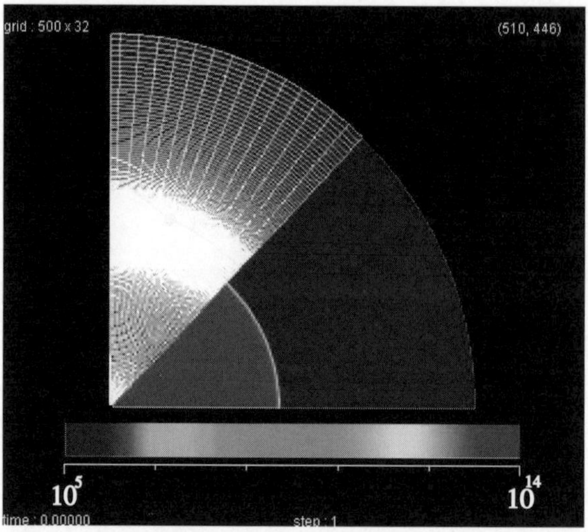

Fig. 3. The nonuniform grids in our calculation and the density profile of the hybrid star. This figure shows the cross section of the star by the polar axis and the horizontal axis. The our grids are shown as white grids in this figure. The color bar means the density profile. The maximum density color is red, which means more than 10^{14} g cm^{-3}. The minimum density color is blue, which means less than 10^5 g cm^{-3}.

3. Result and Discussion

We find that the elements of envelope can eject by "petit" collapse and bounce by the transition. Fig. 4 is one result of our test calculations, and shows 'petit' shock by QCD phase transition. The dynamical time scale of the collapse and bounce are ~ 0.1 milli-second (see the difference of the time between two panels of Fig. 4).

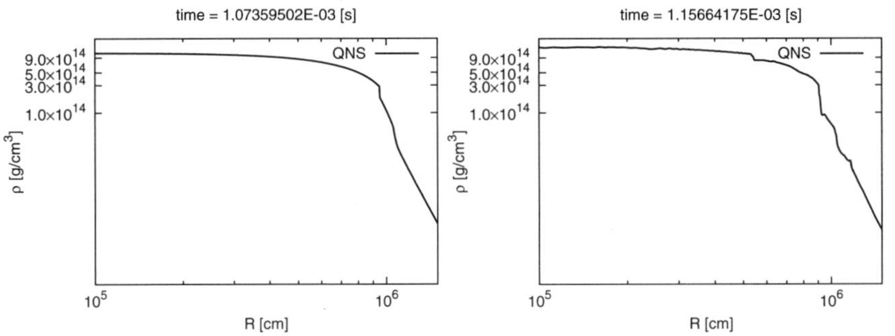

Fig. 4. The shock propagation. (a) Initial stable star. (b) After the transition. We can see 'petit' shock at the outside of the core.

Furthermore we estimate the ejected mass from the envelope. In our simulation, the lowest density (ρ_{min}) of EOS is 5.1×10^5 g cm^{-3} which is the lowest value of the hadron EOS table.[19] Therefore, we can not estimate correctly the physical quantities at the outside of the envelope. We estimate the ejected mass when the density is over ρ_{min} and the radial velocity exceeds sound velocity. This estimation is reasonable, since the we assume that the outside of the envelope is almost vacuum. As the result, the ejected mass is $\sim 10^{-16} M_{\odot}$ for one collapse and bounce (the time scale is ~ 0.1 milli-second). This mass ejection is very little, however not trivial for cosmological time scale, a few billion years.

The temperature of ejected mass is under hydrogen or helium ignition temperature in whole our calculations, since the initial temperature of the envelope is so low, $\sim 10^5$ K. It shows the possibility that the rp-process elements are ejected from the hybrid stars without nucleosynthesis. This possibility have not been suggested ever. In this paper, we calculate spherical model as a test. However, with the rotations or the magnetic fields, it is not clear whether this possibility is correct or not. We will check them at nearly future.

Acknowledgment

N.Y. is grateful to M. Alford for his suggestion, the students in the Waseda University for fruitful discussions. This study was supported in part by the Japan Society for Promotion of Science (JSPS) Research Fellowships (N.Y.), the Front Researcher Program on Kyushu University (T.N.), Grants-in-Aid for the Scientific Research from the Ministry of Education, Science and Culture of Japan (No.S14102004, No.14079202, No.17540267), and Grant-in-Aid for the 21st century COE program "Holistic Research and Education Center for Physics of Self-organizing Systems".

References

1. T. M. Braje and R. W. Romani, *Astrophys. J.* **580**, p. 1043 (2002).
2. J. J. Drake, *Astrophys. J.* **527**, p. L51 (2002).
3. J. A. Pons, F. M. Walter, J. M. Lattimer, R. Neuhauser and P. An, *Astrophys. J.* **564**, p. 981 (2002).
4. F. M. Walter and J. M. Lattimer, *Astrophys. J.* **576**, p. 145 (2002).
5. D. Page, M.Prakash, J. M. Lattimer and Steiner, *Phys. Rev. Lett.* **85**, p. 2048 (2000).
6. D. Blaschke, T.Klahn and D. N. Voskresensky, *Astrophys. J.* **533**, p. 406 (2000).
7. D. Blaschke, H. Grigorian and D. N. Voskresensky, *Astron. & Astrophys.* **368**, p. 561 (2001).
8. H. Grigorian, D. Blaschke and D.N.Voskresensky, *Phys. Rev. C* **71**, p. 045801 (2005).
9. L. M. Lin, K. S. Cheng, M. C. Chu and W. M. Suen, *Astrophys. J.* **639**, p. 382 (2006).
10. N. Yasutake, K. Kotake, M. Hashimoto and S. Yamada, *Phys. Rev. D* **75**, p. 084012 (2007).
11. N. Yasutake, M. Hashimoto and Y. Eriguchi, *Prog. Ther. Phys.* **113**, p. 5 (2005).
12. J. L. Zdunik, M. Bejger, P. Haensel and E. Gourgoulhon, *Astron. & Astrophys.* **465**, p. 533 (2007).
13. M. Fujimoto, T. Hanawa, I. I. Jr. and M. B. Richardson, *Astrophys. J.* **278**, p. 813 (1984).
14. O. Koike, M. Hashimoto, R. Kuromizu and S. Fujimoto, *Astrophys. J.* **603**, p. 242 (2004).
15. R. E. Taam, *Ann. Rev. Nucl. Part.* **35**, p. 1 (1985).
16. W. G. Lewin, J. v. Paradijs and R. E. Taam, *Space Sci. Rev.* **62**, p. 223 (1993).
17. S. E. Woosley, A. Heger, A. Cumming, R. D. Hoffman, J. Pruet, T. Rauscher, J. L. Fisker, H. Schatz, B. A. Brown and M. Wiescher, *Astrophys. J. Suppl.* **151**, p. 75 (2004).
18. M. Shibata, *Phys. Rev. Let.* **96**, p. 031102 (2006).
19. H. Shen, H. Toki, K. Oyamatsu and K. Sumiyoshi, *Nucl. Phys. A* **637**, p. 435 (1998).
20. N. K. Glendenning, *Phys. Rev. D* **46**, p. 4 (1992).

21. J. M. Stone and M. L. Norman, *Astron. & Astrophys. Suppl. Ser.* **80**, p. 753 (1992).
22. A. Marek, H. Dimmelmeier, H.-T. Janka, E. Muller and R. Buras, *Astron. & Astrophys. Suppl. Ser.* **445**, p. 273 (2006).

PRESENTATION TITLES

Table 1. The 5th 21st century COE symposium on Physics of Self-organization System Oral Session Program, September 13, 2007.

Talkers		Presentation Titles
Shuji Hashimoto	Senior Dean, Waseda Univ.	Opening
Etienne Gheeraert	Embassy of France, Attaché	
Shin'ichi Ishiwata	21COE Leader, Waseda Univ.	
Pierre Gaspard	Université Libre de Bruxelles	Thermodynamic time asymmetry and nonequilibrium statistical mechanics
Yoshitsugu Oono	Univ. Illinois	Bio-physics manifesto —for the future of physics and biology—
Bei-Lok B. Hu	Univ. Maryland	New View on Quantum Gravity and Structures of Spacetime: Issues at the micro-macro and quantum-classical interfaces
Shin-ichi Uchida	Univ. Tokyo	How to increase superconducting T_c ?
Hiromichi Nakazato	21COE Member, Waseda Univ.	A measurement-based purification scheme and decoherence
Tsuneaki Daishido	21COE Member, Waseda Univ.	Extragalactic and Galactic Radio Transients with Spatial and Temporal Observations

Table 2. Oral Session Program, September 14, 2007.

Talkers		Presentation Titles
Takaaki Sato	Visiting Lecturer, Waseda Univ.	Protein-Protein Interactions and Their Implication in Medical Application of Artificial Blood Substitutes
Hiroko Yokota	JSPS DC Fellow, Waseda Univ.	Nucleation and time evolution of polar embryos in a quantum relaxor $KTaO_3$ doped with Li
Takaaki Monnai	JSPS PD Fellow, Waseda Univ.	Role of an intermediate state in the Kramers escape rate theory
Yu-ichi Takamizu	JSPS PD Fellow, Waseda Univ.	Colliding branes and its application to string cosmology
Katsuyuki Shiroguchi	Visiting Lecturer, Waseda Univ.	Walking' mechanisms of myosin V, a two-'foot' linear motor protein
Sotaro Uemura	Assistant Prof., Univ. Tokyo.	Single molecule force measurement for protein synthesis on the ribosome
Poster Abstract, PhD Students	Yuta Shimamoto Yuko Urakawa Takehito Suzuki	Soichiro Shibasaki Tomohiro Hasumi Hiroaki Kanzawa
Poster Session		
Concluding and Award	Pierre Gaspard	

Table 3. Poster Session 1.

Names	Presentation Titles
Hisatsugu Yamasaki	Probing Proton Pathways in the Fo Sector of ATP Synthase studied by Molecular Dynamics Simulations
S. L. Lu	Carrier density dependence of nonresonant carrier tunneling in GaAs/AlGaAs double quantum wells-effect of exciton and free carrier thermodynamic
Umpei Miyamoto	Hawking radiation from higher-dimensional black holes and gravitational/gauge anomalies
Takashi Ohki	Single-molecule analysis of myosin IX by optical trap nanometry
Sergey V. Mikhailenko	How load controls the processivity of the oppositely directed motors, myosins V and VI
Yasushi Matsunaga	A new Landau mode in current-carrying carbon nanotubes
Yusuke Oguchi	Angular dependence of ADP dissociation kinetics in myosin V under directional loading
Hiroaki Kanzawa	Variational Calculation for the Equation of State of Nuclear Matter toward Supernova Simulations
Sumiko Kida	Two Strong Radio Bursts at High and Medium Galactic Latitude
Shin-ya Komura	Pattern formation induced by the collective motion of granular particle interacting with the motion of interface
Tomoka Sunaga	Analytic Expression of the Condition for the Existence of Complex Modes for a Trapped Bose-Einstein Condensate with a Highly Quantized Vortex
Yuta Takahashi	Effects of Dispersion on Generation of Squeezed Light
Shigehito Tanahashi	Interaction between circular vection strength and posture
Masaki Nakagawa	Statistical properties of dissipative linear random mapping having infinite invariant density
Junichi Harada	CONSTRUCTION OF NODAL SOLUTIONS TO SOME NONLINEAR ODE'S SYSTEMS
Masato Fujiyoshi	Decay of Unstable State Coupled with a Cantor Set Like Spectrum
Takahito Mitsui	A mathematical model for the intermittent behavior of animal locomotions
Mitsuteru Mimura	Xenon Time Projection Chamber for MeV g-ray Imaging
Daisuke Yamamoto	Pseudospin Representation of Constrained Electron Operators in the Two-Dimensional t-J Model
Yoko Ishii	Non-linear Rheology of Lyotropic Lamellar Phases
Yukinori Sasagawa	Black Hole Solutions in Supergravity with String Corrections

Table 4. Poster Session 2.

Hiroki Nagakura	General Relativistic Hydrodynamical Simulations of the Standing Accretion Shock Instability around a Black Hole: Quasi Periodic Oscillations and their Implications for GRB
Tai Tanaka	A Burst from Radio Transient WJN J0417+4051
Hiroyuki Kubo	Muscle Based Facial Animation using a Few Motion Capture Markers
Syuya Ohta	Measurements of charge changing cross sections for iron nuclei on hydrogen to estimate GCR path length distribution in the galaxy
Tomo Tanaka	Construction of mass operator via loop quantum gravity
Takehito Suzuki	Large Magnetocapacitance of a spinel Mn_3O_4 single crystal
Soichiro Shibasaki	Thermoelectric properties of Ni-doped $LaRhO_3$
Yuta Shimamoto	Self-regulatory mechanisms of force generation in the contractile system of muscle
Yuko Urakawa	Cosmological perturbations in Stochastic gravity
Tomohiro Hasumi	Statistical properties in the spring-block model for earthquakes
Michitaka Katsumura	Temperature and magnetic-field dependence of optical conductivity in spinel MnV_2O_4
Junichi Masuzawa	Temperature-modulated reflectivity measurement for the study of the dynamics of two-phase coexistence
Kohei Azumi	Resistance switching with applied electric field studied by microspectroscopy
Daisuke Satoh	Coupling between itinerant carriers and magnetic impurities in Cr-doped (Sr,La)TiO_3
Haruki Takei	Magnetic-field effect in spinel FeV_2O_4
Nobuhiro Umehara	Molecular Dynamics Study on Ligand Induced Allosteric Transition in Myosin Motor Domain
Emiko Nakamura	Molecular dynamics study on the effect of electrostatic interactions on the F-actin stability
Kohei Meguro	Brain activity relevant to switch of motion perception measured by NearInfrared spectroscopy (NIRS)
Tomohiro Suko	Current Oscillation in the Organic Conductor θ-(BEDT-TTF)$_2$CsZn(SCN)$_4$
Yoshihiro Takahashi	Thermoelectric properties of nano-sized $Ca_{0.96}Sm_{0.04}Mn_{O3}$
Shin Yoshida	Spin-state transition under pressure in $Sr_{0.78-x}Ca_xY_{0.22}CoO_{3-d}$
Naoki Takasugi	Statistical Laws of The Multi-Agent Model for Financial Markets
Kentaro Osawa	Fano effect in a Josephson junction
Ryoko Ichihara	$\pi-$state of Bose-Einstein condensates in a double-well potential

Table 5. Poster Session 3.

Yuki Terakawa	Quantum fluctuations on a bosonic Abrikosov lattice in bilayer system
H. Matsuura	Optical amplified modulation devices using Erbium-doped $LiNbO_3$ crystal
Rai Kou	Nonlinear optical wavelength converter for Si photonics
Jo Hattori	Photo-induced molecular precession in azobenzene liquid crystals
S. Ishizaki	Collective Precession of Chiral Liquid Crystals under Transmembrane Mass Flow
Satoshi Kawado	SHG interference tomography and its application to 3D observations of periodically poled domains
Aoi Okada	Dielectric behavior with two relaxations in quantum relaxor $(KLi)TaO_3$
Makoto Hareyama	Electron burst in the Earth's radiation belt as earthquake precursors observed by Japanese Satellite, SERVIS-1
Naoyuki Yamashita	Search for Water on the Moon by KAGUYA Gamma-Ray Spectrometer
Takuma Akimoto	Road to the Randomness of the Correlation Functions in Infinite Measure Systems
Hiroki Nakamura	Inclusive Pion Production and Quasielastic Interactions in Electron-Nucleus Scattering
Shigeru Ajisaka	Nonequilibrium Peierls transition
Hirotaka Ito	The Evolution of Shocked Shell in Radio Galaxies
Makoto Unoki	Effect of decoherence on a scheme of preparation of quantum states
Kenji Orihashi	Toward the ergodic problem in nonequilibrium stationary system — statistical analysis of Fermi-Pasta-Ulam β model —
Kotaro Niinuma	The Detected High Galactic Radio Transients and The Alert System for Quick Analysis
Yuzuru Karouji	Global Measurement of Chemical Composition of the Moon's crust by Gamma Ray Spectrometer Onboard the SELENE (KAGUYA) and lunar meteorites
Hidetomo Sawai	Magnetized Supernovae and Magnetar Recoils
Takeshi Itabashi	Real-time observation of self-organization processes in the mitotic spindle
Hiroaki Kubota	Mutation of D-loop of actin and its effects on the interaction with myosins, II and V

AUTHOR INDEX